T0222238

Making Changes in STEM Education

Many science, engineering, technology, and math (STEM) faculty wish to make an academic change at the course, department, college, or university level, but they lack the specific tools and training that can help them achieve the changes they desire.

Making Changes in STEM Education: The Change Maker's Toolkit is a practical guide based on academic change research and designed to equip STEM faculty and administrators with the skills necessary to accomplish their academic change goals. Each tool is categorized by a dominant theme in change work, such as opportunities for change, strategic vision, communication, teamwork, stakeholders, and partnerships, and is presented in context by the author, herself a change leader in STEM. In addition, the author provides interviews with STEM faculty and leaders who are engaged in their own change projects, offering additional insight into how the tools can be applied to a variety of educational contexts.

The book is ideal for STEM faculty who are working to change their courses, curricula, departments, and campuses and STEM administrators who lead such change work to support their faculties, as well as graduate students in STEM who plan to enter an academic position upon graduation and expect to work on academic change projects.

Making Changes in STEM Education
Education
The Change Maker's Toolkit

Julia M. Williams

CRC Press
Taylor & Francis Group
Boca Raton New York London

CRC Press is an imprint of the
Taylor & Francis Group, an **informa** business

Cover image: Photo by Daniele Levis Pelusi on Unsplash. Design by Julia Reich.

First edition published 2023
by CRC Press
6000 Broken Sound Parkway NW, Suite 300, Boca Raton, FL 33487-2742

and by CRC Press
4 Park Square, Milton Park, Abingdon, Oxon, OX14 4RN

CRC Press is an imprint of Taylor & Francis Group, LLC

Library of Congress Cataloging-in-Publication Data
Names: Williams, Julia M., author.
Title: Making changes in STEM education : the change maker's toolkit /
Julia M. Williams.
Description: First edition. I Boca Raton : CRC Press, [2023] I Includes
bibliographical references and index.
Identifiers: LCCN 2022055187 I ISBN 9781032390789 (pbk) I ISBN
9781032392554 (hbk) I ISBN 9781003349037 (ebk)
Subjects: LCSH: Technology--Study and teaching. I Science--Study and
teaching.
Classification: LCC T65 .W68 2023 I DDC 607.1--dc23/eng/20230109
LC record available at https://lccn.loc.gov/2022055187

ISBN: 978-1-032-39255-4 (hbk)
ISBN: 978-1-032-39078-9 (pbk)
ISBN: 978-1-003-34903-7 (ebk)

DOI: 10.1201/9781003349037

Typeset in Times
by SPi Technologies India Pvt Ltd (Straive)

Dedication

For the change makers

Contents

Acknowledgments

Of all the challenges I encountered in writing a book about making changes in STEM education, none has seemed as formidable as the writing of these acknowledgments. This book would not be possible without the influence, counsel, and contributions of many gifted colleagues and friends. Although I worry that my thanks will not represent accurately the value of all that they gave to me and to this project, I will try.

Dr. Susan Robison has served as my coach, as well as my example of an author who can write a practical book that supports the work of academics. Susan's skills are extraordinary. She combines wit with insight, so while she helps her clients come to grips with the obstacles that block their path, she is quick to highlight that the obstacles are as likely internal as external. Susan read much of this manuscript and helped transform the text into my readers' book, rather than mine.

Dr. Helen Sword provides useful guidance for academic writers in many forms. For me, the starting point was the 5 Day At Home Writing Retreat, which I then followed with the 30-Day Writing Challenge. Through these and other resources that she shares online and in her books, Helen helped me develop my own writing practice, an essential first step in order to write this guide to academic change. If any portion of the book meets her definition of stylish writing, then Helen deserves the credit.

Ms. Julia Reich, chief easel officer of Stone Soup Creative, provided the change maker illustrations for this book. Julia read each chapter and offered her visual interpretation of my words, further enriching the work. I appreciate her willingness to collaborate on this project.

My perspective on academic change making was nurtured by many skilled change makers in several contexts. Rose-Hulman colleagues and members of the Making Academic Change Happen Workshop—Drs. Eva Andrijcic, Steve Chenoweth, KC Dee, Craig Downing, Richard House, Ella Ingram, Glen Livesay, Matt Lovell, Sriram Mohan, and Don Richards—helped to lay the foundation for this book. Together they are a dream team for productive collaboration. Each of them is a change maker in their own right, and I have learned much from their examples. I also appreciate the willingness of those individuals who decided to attend the MACH workshops from 2012 to 2017 and to cultivate their change making skills in a supportive environment. I hope they are still pursuing academic change on their own campuses and that they will find this book a useful resource.

Drs. Eva Andrijcic, Ella Ingram, and Sriram Mohan have served as co-investigators with me on the Revolutionizing Engineering Departments Participatory Action Research (REDPAR) project funded by the National Science Foundation since 2015. Together, we have enjoyed a productive partnership with Drs. Elizabeth Litzler and Cara Margherio of the Center for Evaluation & Research in STEM Equity at the University of Washington. Through our REDPAR work, we have also benefited from the work of Drs. Kerice Doten-Snitker, Anna Swan, and Selen Guler, who were associated with the project as they completed their degrees. Our practice-research partnership is a powerful example of what can emerge when great people devote themselves to good work. I also wish to thank the Revolutionizing Engineering and Computer Science

Departments (RED) teams who have been dedicated to transforming engineering and computer science education at their respective institutions. Their projects afforded me numerous examples of the challenges that academic change makers face. In many ways, they inspired the creation of many of the tools that you will find in this book.

The Kern Family Foundation supported my sabbatical work on this book. During sabbatical, I collaborated with members of the Kern Entrepreneurial Engineering Network staff in order to support the efforts of engineering faculty in change work. Also during my sabbatical year and through KEEN, I benefited from collaborations with a number of faculty who participated in the KEEN on Stories Project, led by Ms. Janece Shaffer. Working with Janece had a profound impact on both my writing and my teaching, and I would like to thank her specifically for her encouragement.

Throughout this book, I emphasize the importance of working with the right team members. In that regard, I have benefited immeasurably from the experience of working with Drs. Eva Andrijcic and Sriram Mohan. Together we have designed and deployed the change maker curriculum for RED teams, as well as creating change skills workshops for members of KEEN. They are generous colleagues who devote themselves unselfishly to this work.

I interviewed the following change makers for this book: Drs. Stephanie Chasteen, Rachel McCord Ellestad, Elizabeth Housworth, Rohit Kandakatla, Jeremi London, Thor Misko, Khairiyah Modh Yusof, and Tim Wilson. Each of them works to make change in their particular STEM education contexts, and they enriched this book by speaking honestly about the challenges and rewards of making change. I wish to thank them for sharing their unique insights and observations about their change work in the interviews that appear at the end of each chapter.

The late Dr. Jeff Froyd was interviewed in the summer of 2021, and I regret that he didn't get a chance to see this work in print. I hope he would find some value in its contents, although I expect he would, in his unique way, have a number of comments and suggestions that I could apply to subsequent revision. With Jeff's passing, all of us in the field of STEM education lost a leader, a change maker, and a friend.

The late Dr. Bill Kline provided support and encouragement to me and to many colleagues in his various roles on our campus. Sometimes the most influential change makers are the ones who call attention to the work of others, rather than to their own.

As writers, we all need Critical Writing Friends, those individuals who are willing to read draft after draft of a project and offer insightful feedback. For this project, I am grateful for the friends who served in this capacity: Drs. Emma Dosmar and Sid Stamm. Our collaborations have been reciprocal, but I can only hope that I offered feedback to them at the same level that they so generously provided to me during the writing and revising of this book.

Acknowledgments customarily save the best for last, the touching tribute to a spouse or partner without whom the book would not be possible. As it turns out, this book would not exist without the patience of Dr. Nick Williams, who endured my frequent trips to observe academic change in context which helped me formulate the change tools you find in this book. Now that the book is done, I promise that we will be spending more time visiting the historic homes of US presidents, playing golf, watching the Red Sox, and joining our extended family in northern Maine for summers by the lake. Definitely a change for the better!

Author

Julia M. Williams joined the faculty of the Humanities, Social Sciences, and the Arts Department at Rose-Hulman Institute of Technology in 1992, then assumed duties as executive director of the Office of Institutional Research, Planning, and Assessment in 2005. From 2016 to 2019, she served as interim dean of Cross-Cutting Programs and Emerging Opportunities. In this role, she supported the work of faculty who create multidisciplinary learning opportunities for Rose-Hulman students. Beginning in 2012, she served as a founding team member of the Making Academic Change Happen (MACH) Workshop that serves faculty, administrators, and graduate students as they pursue their change goals. Williams' publications on academic change, assessment, engineering and professional communication, tablet PCs, and ungrading have appeared in the *Journal of Engineering Education* and *IEEE Transactions on Professional Communication*, among others. She has been awarded grants from Microsoft, HP, the Engineering Communication Foundation, and National Science Foundation. Currently she is principal investigator on the NSF Revolutionizing Engineering Departments (RED) Participatory Action Research (PAR) project, a practice-research collaboration that provides customized faculty development support for 26 RED project teams. In addition, she works as project director for the Embedding Entrepreneurial Mindset in Engineering Professional Societies (KEEN). She has received numerous awards, including the 2015 Schlesinger Award (IEEE Professional Communication Society) and 2010 Sterling Olmsted Award (ASEE Liberal Education Division).

1 The Change Maker's Toolkit

INTRODUCTION

For many faculty members, graduate students, postdocs, staff members, and academic administrators, the day comes when something must change (see Figure 1.1). It could be the day when the way you teach a particular topic doesn't connect with the current generation of students. Or it could be the moment when department policy no longer serves the needs of department members. Or it could be the realization that if the campus hopes to continue to be vibrant, something must change.

When that moment comes, you may seek out the research on academic change and read about change theory. Unfortunately, theory may not provide you with an adequate guide to make the changes that you see are necessary. What you need instead is a practical guide to academic change that is based on research but provides

FIGURE 1.1 The change maker. (used with permission of the author.)

DOI: 10.1201/9781003349037-1

usable tools that can start you on your change work or help you when you encounter problems. You need a community of change makers who, like you, are striving to make changes, large or small, on their own campuses. You need this book.

Before you read further, take a moment to reflect:

What comes to mind when you think about academic change? Write your reflection here.

- Have you considered initiating a change project, but are not sure how to get started?
- Have you attempted change projects in the past only to be frustrated and beset with obstacles?
- Have you been charged to implement academic change and you want to optimize your effort to increase the likelihood of success?

If so, this book is designed for you, the aspiring or the experienced change maker who is in search of a guide through academic change.

The book is divided into six chapters:

- In Chapter 1, you will be introduced to the state of academic change in STEM higher education and learn how needed change is at this moment;
- In Chapter 2—Getting Started, you will learn about yourself and your potential as a change maker, spend time considering opportunities for change as they currently exist on your campus, and develop a plan for moving forward with change;
- In Chapter 3—Initiating, you will begin to develop your change maker's toolkit, learning about tools that can help you jump-start your change project. These tools are grouped into three primary types: Communication, Teamwork, and Diagnosing Problems. These three tool types will appear in the next two chapters as well and will give you a way to organize your toolkit;
- In Chapter 4—Maturing, you will consider how you can grow and expand your change project. The challenges at this stage are different than they are at the Initiating stage, and the chapter will help you cultivate tools that can help;
- In Chapter 5—Propagating, you will explore opportunities for sending your change idea out into the world, expanding the reach of your change project, and helping others to adapt your project to their own specific academic contexts.
- In Chapter 6, you will take time to consider your development as a change maker and look forward to a bright future.

Because this book is designed as a manual, you can use it in multiple ways. You can start and proceed chapter by chapter if you wish, reading each section of the text and completing all of the exercises and reflections. Or you can use this book by topic, depending on the state of the change effort you are engaged in. For example, if your challenges relate to communication, you can work your way through each of the Communication tools and use them if they seem appropriate. At the end of each chapter, you'll meet academic change makers in STEM who are engaged in important change work. By reading their insights and learning about their projects, you will join a community of change makers who can help you as you pursue your change goals.

Of course, you may have immediate need for the content in this book and can't take the time to read through it from start to finish. You can find what you need by digging into specific chapters:

- For example, want to brainstorm ideas for change and formulate them into a project? See Chapter 2.
- Wondering how you can influence others on your campus to look upon your project with favor and interest? See Chapter 3.
- Having difficulties with team members? See Chapter 4.

With each change maker tool you acquire, you will adapt the tool to your specific campus context. Each campus is different. The values of each department are different. The individuals you have to work with are different. This book assumes that you know your own context best and will adapt each tool to your particular circumstances. This book works best when you take what it offers and make it your own.

You may be eager to find answers to your questions right away, so moving into specific sections of the book may be your best strategy. On the other hand, as a change maker, you may benefit from learning more about the need for change within STEM higher education.

ACADEMIC CHANGE IN STEM HIGHER EDUCATION

Consider the following change scenarios that can be found in many STEM educational settings. You may find one or two to be familiar, or you may have experience with another that fits the same formula:

1. A tenure-track faculty member is assigned to teach a standard required course within her department during her first year on the job. She is given a textbook and course notes from a colleague, but she soon finds that teaching by following the proscribed methods isn't engaging her students.
2. A faculty member is hired into a department and charged to develop and deploy a new curricular program. From the start, the faculty member encounters strong resistance to the program, and a senior colleague suggests that the faculty member would be better off (and more likely to be retained) if they slowed progress on the program.

3. A multidisciplinary team is awarded a large external grant to develop a new curriculum that promises to impact student outcomes positively. There is clear support for the effort from administrators, employers, and other stakeholders. Once the team seeks others to join the effort, however, they have difficulty finding additional individuals who wish to join them in the work.

4. A college dean is nationally recognized as a change leader by their peers. This individual has gained support for a new academic initiative and has proceeded to enlist academic personnel to join in. But as the project encounters stumbling blocks along the way, the dean pushes forward, ignoring input and concerns from those individuals who were early adopters. In relentless pursuit of the project objectives, the dean burns bridges with faculty, academic staff, department chairs, and administrators.

5. A team of faculty, professional staff, and administrators embarks on an important collaborative project with another institution. Soon after the project begins, several of the original leaders of the project leave to pursue other opportunities. The remaining members of the team are uncertain regarding who the project's leader is.

Unfortunately, the landscape of STEM higher education is littered with projects like these: projects that began with a solid foundation but fail to progress and thus stand half-finished; faculty, staff, and administrators who are willing to embark on innovative new academic initiatives only to find themselves without direction or leadership, mystified about how this happened. And when change projects fail, often we hear the same rationales for their failures: "faculty resist change," or "change in academia is hard," even "change isn't necessary in higher education."

For the moment, let's set aside these conventional rationales for failed change in higher education and consider instead the change makers themselves. Each change project, whether it succeeds or fails, begins with individual change makers, those who envision improvements (whether in faculty teaching, student learning, or some other aspect of higher education) and seek ways to implement change. These people are willing to invest time, energy, and social capital to make those improvements a reality. You might expect that those who succeed in making academic change happen are themselves in possession of significant and necessary attributes that help them be successful, like a secret formula for change. Little attention has so far been paid to isolating these attributes and understanding how they contribute to the success of a change project. To be a successful change maker, an individual should be in possession of a change maker's toolkit, the set of "tools"—also called competencies, skills, abilities, and knowledge—that allow the change maker to initiate a change project and convince others to follow.

Referring to these attributes as a "toolkit" may bring to mind a carpenter's toolbox, with everything necessary to cut a 2x4 or nail down a floor. In general, each profession requires its own set of specialized tools, and the process of acculturation into the profession—whether you are a carpenter, engineer, nurse, scientist, electrician, accountant, social scientist, humanist, or any other professional—is a process of learning to use the tools well. Like any skilled professional, a change maker needs a toolkit, a specialized set of knowledge, skills, and abilities that can help the change maker start a movement, craft a vision, help team members reach their potential, accomplish goals, and essentially change the landscape for higher education.

The question is, how do we know that change makers, like individuals in other professions, need a specialized set of competencies in order to initiate, lead, and achieve change? And if we believe that these tools are needed, which competencies, which "tools," are the right ones and how do we know if they work?

Fortunately, the research literature on the study of change offers insights into these questions. We can, for example, look to science education research, particularly the work of Charles Henderson, Andrea Beach, and Noah Finklestein (2011). They proposed a four-quadrant analysis of change as it was represented in 191 conceptual and empirical journal articles published between 1995 and 2008. From this literature review, they created four broad categories of change strategies—Disseminating Curriculum & Pedagogy, Developing Reflective Teachers, Developing Policy, Developing Vision—to capture the core differences within the body of literature that they analyzed (see Figure 1.2).

Each change category maps to specific change strategies appropriate to it. For example, creating a learning community (a strategy associated with the Reflective Teachers category) is well aligned with growing faculty knowledge about effective teaching practices. In contrast, imposing a department policy (a strategy associated with the Policy category) is not well aligned with that same change. Thus, these

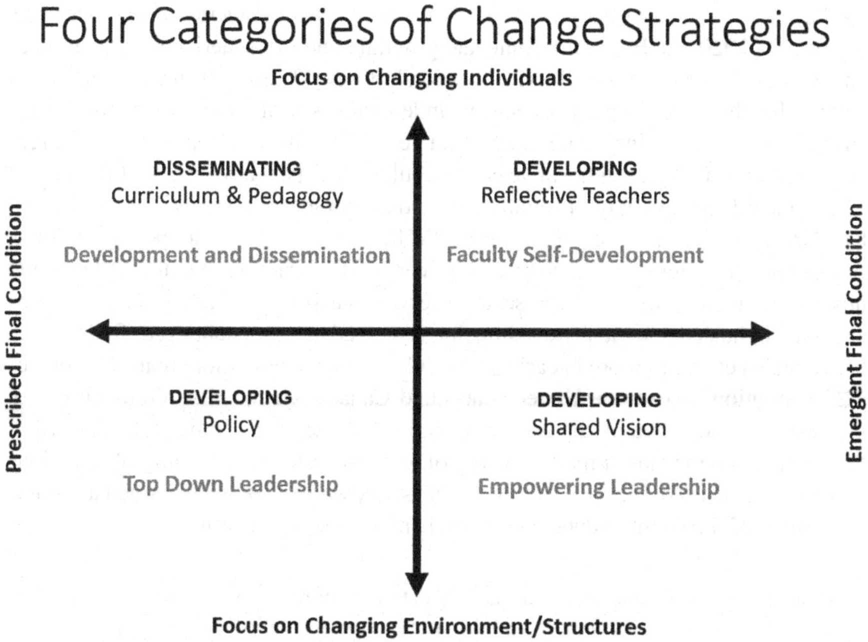

Four Categories of Change Strategies

FIGURE 1.2 C. Henderson, A. Beach, and N. Finkelstein, "Facilitating change in undergraduate STEM instructional practices: An analytic review of the literature". *Journal of Research in Science Teaching*, 48(8): 952–984 (2011). (Used with permission from the publisher.)

authors argue, change has been difficult to achieve because change makers do not match the strategy to the change they envision.

The authors' analysis leads to an important point. It is a frequently cited assumption in higher education that people and systems resist change. According to Henderson et al., resistance in higher education is a symptom of the use of a change strategy poorly aligned with the objective.

- Change makers may not be able to gain buy-in for their project from individuals.
- Change makers may have difficulty aligning a vision of change among multiple individuals and across multiple departments who must collaborate to make the change happen.
- They may hear faculty complain that the proposed change threatens their academic freedom, their sovereignty in the classroom, and even the future of the institution itself.

Addressing these and other issues may require skills that are not part of the change maker's disciplinary expertise. The ability to make change requires a specialized set of competencies. Beyond disciplinary expertise, change makers need a toolkit, a set of competencies that can prepare the change maker to operate more effectively in their academic environment.The time for addressing change in STEM higher education is pressing. Across the United States and abroad, there are repeated calls for changing the way we educate our students, such as the National Academy of Engineering's Engineer of 2020, the Educate to Innovate program, and the American Association of Universities' Undergraduate STEM Education Initiative. These calls for change extend beyond the classroom experience and include changes to the curriculum, to the co-curriculum, and to the institution itself. And yet, despite the development of research-based teaching strategies, innovative co-curricular projects, many years of funding and development from a variety of foundations and corporations, even the conditions produced by a global pandemic, change in STEM higher education is not pervasive. Take, for example, the attempted move to student-centered instruction in many colleges and universities. Results from a large-scale observational study of undergraduate STEM education indicated that faculty teaching has remained largely unchanged (Stains et al., 2018). Stains et al. monitored nearly 550 faculty as they taught more than 700 courses at 25 institutions across the United States and Canada. Of the classrooms observed, 55% were characterized by the instructor using a "didactic" teaching style, defined as 80% or more of the class time consisting of lecturing. Of the remaining observations made, 18% consisted of a "student-centered" style, while 27% were defined as "interactive lecture." From these data, the researchers expressed concern:

> Although we are unable to claim that our data are entirely representative, the sample size and diversity of courses and disciplines represented in our data suggest that these profiles and broad instructional styles provide a reliable snapshot of the current instructional landscape in undergraduate STEM courses taught at North American institutions.

(1469)

Take a moment to consider your own classroom, department, college, and campus. Where does your institution place, in the 55%, or in the 18%?

The lack of systemic change points to an important problem with the approach to change that the STEM higher education community has pursued thus far. Change has been targeted at the course and curriculum levels (Borrego et al., 2010), but course-level changes have not led to systemic change across departments and institutions. Such failure suggests that our focus has been misplaced. We have largely ignored the individual development dimension, which would require preparing faculty, academic staff, graduate students, and administrators with the necessary knowledge, skills, and abilities (KSAs) in motivation, communication, collaboration, and persuasion. These change strategies are well documented in the literature of disciplines such as organizational psychology and behavior (e.g., Daly and Finnegan, 2010; Quinn 2010), but have not been brought into the conversation within STEM higher education in a rigorous, accessible way. This book attempts to address this gap.

Making Changes in STEM Education: The Change Maker's Toolkit addresses this specific problem by posing the following question:

> Can we overcome limits that prevent the diffusion of new ideas, can we overcome barriers to the adoption of effective practices, by focusing on the change agents themselves in terms of their skills and change expertise?

In order to equip individuals with the right tools, we must acknowledge that most faculty in the science, technology, engineering, and math fields are disciplinary experts and not familiar with the research literature of change. Within the literature of higher education, organizational psychology, and other research areas, there are theories of change that are useful (Eckel et al., 1999). If, however, a faculty member outside of the disciplinary area of higher education research seeks out change maker tools, they must know where to look and how to translate the research findings into usable, practical applications.

> What is needed to prepare and support change makers in STEM is a book that translates the research literature on change into specific practices that faculty, staff, graduate students, and administrators can use, whatever their institutional context or planned change project. *Making Changes in STEM Education: The Change Maker's Toolkit* is that book.

In this book, the model for change is based on the faculty development model, a model that is widely adopted on campuses across the United States and abroad, but has been seldom applied to the context of preparing faculty to make academic change. In this model, we start with the individual—a professor, a graduate student, an academic staff member, or an academic administrator. This person recognizes a gap, a problem, and/or an opportunity that they could remedy with a new way to do things. In order to resolve the problem, the individual may be able to rely on their current state of knowledge and skills, but it is more likely that the opportunity requires something more from the individual who pursues it.

Think of this in the case of the faculty member who wishes to introduce a new teaching method to an existing course. She sees that her current pedagogy is inadequate to address students' needs in her course. So she seeks out new knowledge, applies for resources from her department head or dean, confers with colleagues who have similar concerns, and makes use of the resources available through her campus teaching and learning center. All of the faculty development support helps her expand her knowledge, skills, and abilities beyond her initial training as an expert in a single academic discipline. We don't expect, for example, a PhD in chemistry to pursue a second degree in curriculum and instruction, but we do encourage the PhD to attend faculty development seminars, consult with academic staff at the center, and collaborate with colleagues who are also engaged with change in their respective domains. For their part, colleges and universities create the teaching and learning centers, and hire experts in the field of faculty development, all because the research in the field of pedagogy points to the efficacy of providing this support in order to help faculty improve their teaching so students can improve their learning.

Similarly, the faculty development model—providing the individual with resources and experts who can support her efforts—works well for the individual who wishes to make a change in their immediate circle of influence, that is, in their own course. The challenges come when the individual decides that the change they seek has application beyond the classroom over which they have nearly absolute control.

- Perhaps the course that the individual teaches is also taught by colleagues in the same department. Do they also see the gap that inspired the faculty member to pursue a remedy?
- The department chair identifies the gap in the current curriculum that impacts teaching and learning not only for majors but for students majoring in other departments. Can the department chair promote change in departments for which they have no direct responsibility?
- The graduate student sees the negative impact that a university-wide requirement has not only on their own progress to degree but on many other graduate students as well. What steps can the graduate student take to remedy the situation?

Unlike the change that can be implemented by the individual alone, these examples illustrate that the individual needs to develop additional skills and employ tactics that go beyond the support the faculty development center can provide. The model of faculty development is applicable here, however, even if the set of skills differs from the customary focus on pedagogy and curriculum development. Individuals may need to develop new skills and acquire new knowledge which, like the help directed toward pedagogy and curriculum development, lie outside of their disciplinary domain.

This book adopts the faculty development model from pedagogy and curriculum improvement and reworks it for the pursuit of academic change. Like pedagogical improvement, change skills and knowledge can be learned. Faculty, graduate students, academic staff, postdocs, and academic leaders don't need a second PhD. They need:

- Support, guidance, information, and expertise.
- Community consisting of fellow change makers with whom to collaborate and confer (whether those individuals are on their home campus or elsewhere).
- Practice in order to acquire new skills.

MY EXPERIENCE AS AN ACADEMIC CHANGE MAKER

The faculty development model applied to academic change derives from my own experience and from the experiences of others I have worked with in numerous STEM higher education contexts. My work in academic change began with my own efforts to make change happen on my campus. Early in my career, I worked to effect change, from advocating for the use of laptop computers in the composition classrooms at Rose-Hulman in 1996 (when we adopted a campus-wide 1:1 laptop program) to developing a co-curricular leadership academy for undergraduate STEM students in 2008. In these and other instances, I was able to identify a need for change, but I often struggled with employing effective strategies that could help me make the change that I wanted to see. I often felt thwarted by the resistance of other faculty or administrators. I believed that the change I envisioned was crystal clear, but I found that I had difficulty communicating what I wished to change successfully enough to convince others.

In 2012, I collaborated with other colleagues at Rose-Hulman who were also engaged with important change projects and who confronted some of the same challenges. Once we surveyed the research literature on change and began to translate its findings into practical strategies, we thought we could share these strategies with other academics. In 2012, we offered the first Making Academic Change Happen (MACH) workshop on our campus. That initiative grew over the years and was the foundation for the work I started with the National Science Foundation and the Revolutionizing Engineering Departments (RED) program in 2015. I now work to change STEM education by supporting collaborations with faculty, academic staff, graduate students, and administrators across the United States and abroad who are engaged in similar efforts.

This book operates on a couple of basic principles. First, it is important to understand who you are as a change maker, what motivates you to do this work, and what your skill set is at the start. In other words, much of what you do as a change maker starts with self-awareness. If you don't start there, you can't understand those around you, people who could either join you in your effort or stand in your way. You can't develop empathy for the situation of others who might feel threatened by your change project if you don't develop the ability to see the situation you wish to change from their point of view.

Second, change makers need a community. Change makers often feel isolated on their campuses when they embark on a change project, and they seek a community of change makers with whom they can learn best practices, share resources, and seek advice. Change makers working in a variety of contexts have encountered many of the same difficulties, and community support can highlight the variety of change strategies that can be enlisted.

Third, change makers need practical tools that are based on academic research literature but are easily implemented. If an individual encounters a challenge in pursuing a change project, they need a readily accessible resource that can offer help.

Making Changes in STEM Education: The Change Maker's Toolkit offers its readers the tools, the community of like-minded change colleagues, and the practical advice they need to accomplish their change objectives whether large or small. At this point, you may be eager to begin to fill your change maker toolkit. Our first step, however, is to gather a few insights regarding the challenges of academic change, presented in the form of a fictional Dear Colleague letter.

REFERENCES

Borrego, M., J.E. Froyd, and T.S. Hall. 2010. Diffusion of engineering education innovations: A survey of awareness and adoption rates in U.S. engineering departments. *Journal of Engineering Education* 99(3): 185–207.

Daly, A.J., and K.S. Finnegan. 2010. A bridge between worlds: Understanding network structure to understand change strategy. *Journal of Educational Change* 11: 111–138.

Eckel, P., M. Green, B. Hill, and W. Mallon. 1999. *On Change III: Taking Charge of Change: A Primer for Colleges and Universities*. An Occasional Paper Series of the ACE Project on Leadership and Institutional Transformation.

Henderson, C., A. Beach, and N. Finkelstein. 2011. Facilitating change in undergraduate STEM instructional practices: An analytic review of the literature. *Journal of Research in Science Teaching*, 48(8): 952–984.

Quinn, R.E. 2010. *Deep Change: Discovering the Leader Within*. San Francisco: Jossey-Bass.

Stains, M., J. Harshman, M.K. Barker, S.V. Chasteen, R. Cole, S.E. DeChennePeters, M.K. Eagan Jr., J.M. Esson, J.K. Knight, F.A. Laski, M. Levis-Fitzgerald, C.J. Lee, S.M. Lo, L.M. McDonnell, T.A. McKay, N. Michelotti, A. Musgrove, M.S. Palmer, K.M. Plank, T.M. Rodela, E.R. Sanders, N.G. Schimpf, P.M. Schulte, M.K. Smith, M. Stetzer, B. Van Valkenburgh, E. Vinson, L.K. Weir, P.J. Wendel, L.B. Wheeler, and A.M. Young. 2018. Anatomy of STEM teaching in North American universities. *Science*, 359(6383): 1468–1470.

DEAR COLLEAGUE LETTER (A FICTION THAT MAY BE CLOSE TO REALITY)

Dear Colleague,

I thought I would reach out to you with this letter, following up on the department meeting yesterday. During your presentation about your curriculum change proposal, I saw how enthusiastic you are about the development work you and your team have done. You have data to support the proposal, showing the positive impact the change will have on student learning and faculty teaching. I was persuaded by your evidence and by the fact that I know you are a dedicated colleague who wants the best for the department.

It was clear, however, that you encountered opposition to your proposal from some faculty and heard no support voiced by others who you believed are in your corner. The vocal opposition raised a series of what I thought were baseless arguments in opposition to the curricular change, using phrases like "We tried that in 1982. It didn't work then and it won't work now!" or "You obviously don't understand how we do things here at the U." In the face of those arguments, you offered more data charts and more references to authorities and research articles, a good strategy to refute objectives during an academic conference, but one that didn't seem to convince the naysayers in the room. And perhaps you hoped to hear from other colleagues who had encouraged you to pursue this project, who were willing to be supportive during one-on-one office chats, but wouldn't speak up in the meeting. I won't claim to know what you were thinking or feeling when the meeting ended, but if I were in your shoes, I would feel disappointed, frustrated, and unsure whether it is worth going forward with the proposal.

I want to take this moment to say, don't quit, don't give up, don't let this setback dissuade you from continuing to work for what is right. Let me suggest, however, that in order to overcome the obstacles that are in your path, you need a different set of tools to make that path smoother. These tools are not the ones you learned as you became a disciplinary expert. Most of your colleagues in the department don't know these tools, but they could make a difference for you and the changes you wish to make.

How do I know about these tools? How do I know that they work? Because I learned them by trial and error. I call them change maker tools, and by adopting the *Making Changes in STEM Education: The Change Maker's Toolkit*, I believe you can equip yourself so that you can handle every naysayer, departmental discussion, presentation to the dean or provost, and implement your proposal with the support of your colleagues.

Learning how to use these tools isn't particularly difficult. It helps to have a community of change makers, people like you who see what needs to be changed in academia and want to work to make it better. The community has expertise and experience, and they want to share it with you.

I apologize for going on this long. I have put a copy of *Making Changes in STEM Education: The Change Maker's Toolkit* along with this letter in your department mailbox. I hope you will take a moment to learn about the tools and practice them, with me and with others. The book could be helpful for members of your committee/team who can serve as advocates for the proposal. Just know that the work you want to do is important and your proposal should be adopted. So let's make that happen!

FOUR PERSPECTIVES ON ACADEMIC CHANGE

Each chapter in this book concludes with an interview with an academic change maker. To conclude Chapter 1, you will meet four academic change makers drawn from four different academic disciplines: physics, mathematics, engineering education, and engineering. In a series of interviews I conducted from 2019 to 2022, they discussed their perspectives on their work to make change on their campuses, and their insights can provide you with important perspectives on making changes in STEM education.

As they reflected on their work, each of them offered a brief and memorable insight about academic change:

- Stephanie Chasteen: "You want to support people in going where they should go and going in a productive direction based off of where they are. You want the ideas to come from them, and sometimes you might find ways to plant those ideas then nurture them."
- Elizabeth Housworth: "I hope that I made no big change. Instead I did 1,000,001 little tiny things that made things better, 1,000,001. Just a vast list of tiny things that made the world better."
- Rohit Kandakatla: "We always work through people. We have to appreciate them, that is what is important in order to lead true change because change always has to come from within."
- Tim Wilson: "One thing I think we do as change agents in academia is we latch onto the formalities of change, like the formal proposal document. Nobody wants to read about the change through the proposal. You have to brief people about the change through informal, non-structured ways. Find ways that they think they can improve it. Ask for their advice. All those classic things about how to get buy-in from people. That's what is valuable."

From reading their interviews, you'll hear four different voices discussing their change work in context, and hopefully you will be inspired to begin your own change work, or view your current change work in a new way, because of their insights.

Julia Williams (JW): **Let's start with telling me who you are, where you are, what your role and/or roles are.**

Stephanie Chasteen (SC): I'm a physicist working in educational change and education research. I spent about 15 years at Colorado University Boulder so that role is most relevant to the topic of your book. I was a postdoc in the Science Education Initiative (SEI) in the physics department at CU Boulder and then continued on at CU Boulder working as a research associate there both doing professional development for faculty and disseminating the results of the Science Education Initiative. Eventually I was the associate director of the Science Education Initiative there. But I now have one of those tenuous positions; formally I'm a "person of interest" at CU Boulder, which means I don't actually even have a position, I just have library access. So, I'm now completely independent. My consulting work is certainly in the same Venn diagram as the SEI stuff. I'm an independent consultant doing external evaluation of large NSF STEM education grants, mostly dealing with departmental and institutional change and faculty development but it's not like I'm particularly using the SEI model. I'd say there's just a different set of change agent things that I'm doing there.

Elizabeth Housworth (EH): I'm Elizabeth Housworth, and I am the former chair of the Department of Mathematics at Indiana University Bloomington and the current chair of the Department of Statistics at Indiana University Bloomington. I got my PhD degree under Loren Pitt at the University of Virginia. And my thesis was in relatively pure mathematics, the intersection of harmonic analysis and probability theory.

Rohit Kandakatla (RK): I graduated from Purdue University with a PhD in Engineering Education in 2019. After graduation, I moved back to India to my hometown in Hyderabad, which is in the south of India. Currently I'm serving as a director for Strategy, Operations and Human Resource Development at KG Reddy College of Engineering and Technology, and these are the three areas in which I've worked for the last three years. My position is mainly administrative, but I do a lot of engineering education research focusing on improving the practice in both my institution and another institution. I also have a part-time role as an adjunct faculty at the Center for Engineering Education Research at KLE Technological University. It's the university that has started the first PhD in engineering education in India. I'm working with them to help them build the program and also drive the research in the Center.

Tim Wilson (TW): I'm in Port Orange, Florida, which is just south of Daytona Beach. I'm currently a beach bum. That's my official title. Well actually, I'm trying to start up a business, and I'm working on a technical research project, because when I was a faculty member and a department chair, I never had that opportunity. Once I retired, I had this stuff that I'd been wanting to work on and so I said to myself, "Okay, I have the time I need now, so I better do something." And I've been working on this since. But before that I was the department chair at Embry-Riddle Aeronautical University in the department of electrical engineering and computer science. I was there for 21 years, serving as chair for about ten years. Before that, I was at the University of Memphis in an electrical engineering department that was combined with the computer engineering program at that time.

JW: **My first question relates to identifying the need for change in your specific academic context. How did you know that change was needed?**

SC: I was involved in the Science Education Initiative that was trying to support change in STEM departments. Its purpose was to address a couple of the things that faculty tend to struggle with when they're making changes to their department or to their courses, such as having time to make those changes and having the expertise in education to make those changes. So, the SEI addresses that problem by hiring postdocs with expertise in the discipline to partner with faculty. Those postdocs either had an interest and background in education or they were trained in education and education research by the program. All of them got training but some of them came in with more or less experience prior to being a postdoc. They really acted as coaches, mentors, and partners with the faculty in transforming their courses. The idea was that it reduced some of the burden on the faculty member.

The expressed purpose was to transform courses, but in the end, faculty themselves were changed. It was really faculty development in the guise of course transformation. It was supposed to be about the change and not as much necessarily about the courses but some of our published work also shows that there were large scale changes to the courses themselves, especially at University of British Columbia, the sister institution. We determined a percentage of student credit hours that were impacted at UBC, and it was a majority of the credit hours in the department. At this scale, you can imagine that the department can really start to have its culture change, and the norms of teaching change.

The idea with the SEI was that you would have two to three postdocs at any one time in the department, maybe overlapping and having different lengths of service. And then, you end up with the post docs impacting pretty much every faculty member or many faculty members in the department, and you could get that culture change. But for other people who have taken on the SEI model, they see that it's really expensive so a lot of people have said, "Well, can we hire one postdoc to work with a few faculty here and there?" And I'd say the jury is still out as to whether or not that can happen, and can it impact cultural change.

EH: When I arrived in the Math Department at Indiana University, I did not particularly want to effect any change besides getting another NSF grant and becoming full professor. The Math Department at Indiana University does not want people going in and coming to us to make large changes. In fact, we have been skeptical when the department has wanted to hire somebody who seems to have strong administrative ambitions, because we don't want somebody coming in and using our department for their personal ambitions to become an administrator.

It is true that I came in after I left a previous department with a boatload of problems. So I came into another department, and math departments universally have problems. They are universally, and more so at that time, but even still, universally male. I mean, it would be hard to find a Research Tier 1, Research 1 mathematics department that had a substantial number of serious women mathematicians. Some departments get around it a bit by combining math education and then the women tend to be in math education or something like that. But you just don't have a lot of R1 institutions with a lot of research active women mathematicians.

It wasn't that I was expecting things to be roses, but I expected it to be a lot better. The reason it was a lot better was because when the chair hired me, he told me that the vote was unanimous. So, I felt supported. The faculty could be as sexist as they wanted to be, but I felt supported. Indeed, they were sexist, but if you're going to survive, you have to brush a lot of that stuff off.

So, I did not come in wanting to effect change in the department. What I did was I came in and I did the same thing that I did at my previous institution: I observed. I got involved in a lot of different activities in different realms on campus. I learned a lot, not only about the Department of Mathematics, but also about how the university worked. Before I became chair, I was on hiring committees for interdisciplinary activities. I was chair of the mathematical modeling subcommittee of the General Education Committee when the campus was starting to implement general education. I served on the College Undergraduate Education Committee. I served on the Budgetary Affairs Committee, which advises the provost on the use of the provost fund and used to consult with the university on larger budgetary issues.

Before becoming chair, I served on the College Tenure Committee. So, I saw a lot of tenure cases from the college. I'm sure I did other things that I'm forgetting, but I served on a lot of committees in a lot of different areas. What I noticed was that I often had the best ideas. In the College Education Committee, for example, the issue was whether the Comparative Literature department should scale back the number of semesters of foreign language it required for its undergraduates. They wanted to do that, because they were under pressure to increase majors. So, decreasing requirements is a way of making your major a little bit more accessible. But this was comparative literature. Comparative literature practically by definition is comparing across languages.

So, I was the one who asked, "What are the requirements at any other department of comparative literature for their undergraduates?" I was the one who asked that question. It wasn't the current undergraduate dean. It wasn't anyone closer to the field. It was me.

RH: The Indian educational system is quite different from yours in the US. As a result, we don't have flexibility in terms of changing the curriculum. We just have to follow what's given to us. We also don't even have flexibility in terms of the assessments. The assessments are conducted externally, and our students actually go to a different institution and write the exam. It's structured that way to ensure there's no malpractices by the institution in terms of the examination. In Indian higher education, at least in the tier two and tier three institutions, there's a lot of concerns about malpractice and lack of quality of education, which is why there are lot of stringent measures from the regulatory authorities to ensure that majority of institutions can actually do good work and the quality of education remains high.

But while this was put in place long time ago, I think we are now coming to a place where these structures and regulatory measures are actually impeding our growth because the university education system is very old and not much has changed. Yes, institutions have changed the curriculum, but the institutional structure has not been changed in the last few decades.

So currently, as we try to innovate and adapt to the industry needs, there's a constant push from the university asking us not to really deviate from the prescribed curriculum. While it started as a way to keep a check on quality, now it's actually created barriers in terms of growth and other aspects. Recently in 2020, however, India announced the National Education Policy, which recommends radical, post-colonialism changes in terms of our education system. We inherited the higher education system from the British, and it was designed to build a class of people who can serve the country. There's no integration of critical thinking or problem solving. There is no expectation that people ask questions about what they are doing and why. In that regard, the system is quite different from the US. Now, however, India is radically transforming its systems by recommending that the regulatory barriers go away.

TW: Oh, I don't know. I just want to make things better. Isn't that what we're supposed to be doing, trying to make things better? And I was never bashful about doing that. I went back to visit some profs at MIT a couple years after I started teaching, and I met up with Tom Weiss, who was on my doctoral committee and who I had as a recitation instructor when I was a freshman way back in 1974. We reconnected, and he was one of my mentors. Tom said, "Hey, I think you'll like this. You like teaching, I think you'll love this," and he hands me a book on pedagogy. I'm like, "Yeah, I like this. You're damn right I like this. This is some good stuff because students are engaged and they're talking."

My thinking about teaching comes out of the EE and CS programs at MIT. You have lectures, you have recitations, then you have tutorials, where the grad student and four or five students work problems at a white board. That's really where the learning happens. That's what I learned with Weiss. One week we swapped roles. He said, "You teach the sections, and I'll do the tutorials." Because he wanted me to get the experience.

And so, he'd come back to the tutorial, I said, "Well, what'd you do?" He says, "I just gave them a bunch of problems to work. I just sent them to the board and had them work problems." And that's when the penny dropped for me, that's when I got it. They need to spend time working problems, doing stuff. It's not me in front of them. After that, I tried to incorporate more and more of that into my instruction as time went on. And so, eventually, I was doing almost a purely flipped class, at least for circuits. I did peer instruction in a couple of classes, and students liked it. You're talking to them. You're asking them questions. They're given permission to engage and say things and be stupid every now and then and have fun. I think the first rule of teaching is, don't be a jerk, right?

And I never got bad student evaluations. I was fair. I wasn't dicking students around. And I would even do stuff like call students on plagiarism. Other faculty were like, "Oh, no, don't do that. You're going to get a lawsuit." But I thought, Well, so what? They cheated. What do you want me to do, say they didn't cheat when they did? I'm trying to get them to stop doing it. They're not bad. They just screwed up. People just make mistakes.

When I got to Embry Riddle on a one year visiting appointment, the department asked me to "Do ABET." And so, they gave me some things to do where I could apply my skills at organization and my energy. And so I did ABET, and I did a bunch of stuff. But then I wasn't going to move forward with a career in academic leadership, so I went the route of Faculty Senate. I was the speaker of our faculty, and then I became department chair when they merged the departments of electrical and computer engineering. I was brought in for the merge, and I had to manage a successful merge, but I could never get these people to really buy in, except for people who are already intrinsically interested in instruction. People who are not interested in improving their instruction are not interested in improving their instruction. They're going to go in the classroom and do what they can get away with. When people are interested in teaching, they will attend a workshop and work to improve how they teach.

JW: **How did you implement change and make it stick?**

SC: I'll preface this by saying my experience is mostly at CU Boulder. At that time I worked closely with Warren Code at University of British Columbia. He is the associate director and continues to direct their education transformation efforts there. He's a mathematician. Despite the disciplinary differences, he and I are very similar. I came in as a postdoc in physics; he came in as a postdoc in math. We both became associate directors, but while I transitioned out of that, he stayed there and continues to coach and train all of the new teaching and learning fellows at UBC. And he is the co-author with me on the SEI Handbook.

When I came into my position as associate director, there were no more new teaching fellows, so I wasn't really involved as much in training and coaching them, so everything I'm about to tell you, I'd say I'm putting on my Warren Code hat from when we wrote the handbook because he had a much broader perspective. So, time and expertise, for sure, are one. Another is that it's hard. Well, a challenge for the postdocs was how to work with the faculty in a way that is supportive of those faculty interests rather than pushing change on them, trying to meet faculty where they are and be a coach and a support rather than trying to push them somewhere that they didn't necessarily want to go. And honestly, it's a lot like parenting, right? Not in a derogatory way, I want to stress, but you want to support people in going where they should go and going in a productive direction based off of where they are. You want the ideas to come from them, and sometimes, you know what ideas you want them to have and so you might find ways to plant those ideas. So, Warren's phrase was "plant seeds and nurture sprouts," which I love. Faculty are very interested in student voices and so we stressed listening to the students and then bubbling up some of those ideas that came from the students. For example, a student could give feedback such as, "Oh, I didn't really

understand what was happening by minute 15 in that lecture." Sharing
that with the faculty member would be an example of planting a seed.

And then nurturing the sprouts, finding those faculty in the department
who are interested in exploring and helping them to go a little bit further.
We screened in the hiring of new teaching and learning fellows by asking
them this question: "What would you do if a faculty member wasn't interested in changing their course?" We thought it counted against the candidate if they said something like, "Well, I would show them a bunch of data
and explain to them why teaching interactively is better." So, that was one
thing that's challenging but also faculty are within the culture of their
department. Faculty could experience subtle pressure from other faculty
by virtue of conversations going on around them about teaching. For
example, there's often this weird unspoken rule that you don't watch
somebody else teach. I think it's department specific and institution specific. As part of the program, we would encourage faculty to go and watch
each other's course because it just gives you ideas when you watch someone else teach, but it was like pulling teeth. I would have to explain to
Professor Y, "We are going to go watch Professor X's course at 11:00 a.m.
on Tuesday," then I would have to go to Professor Y's office and walk
them over. They just needed that permission and that prompt to go but it
was always valuable. So that's definitely a challenge. If you could find a
student who was saying exactly what you, as the teaching and learning
fellow, were observing, then you could share something very powerful
with the faculty member. So, a lot of the data that we would collect would
be student information, in a wide range of forms, everything from casual
observation to systematic interviews and concept inventory development.

EH: As a member of the college General Education Committee, I wrote most of the
mathematical modeling language and dealt with disputes when departments outside of mathematics wanted to offer mathematical modeling
that had to be negotiated. That issue was part of the story of how the
Statistics Department at IU was formed in 2006. Two statisticians in the
Math Department were essentially bribed to move to the Department of
Statistics. They didn't want to, but they were offered substantial pay
increases for moving. I was also offered a pay increase to move, but mine
was smaller, and I said no. This is actually a very good move on my part.
At the time, I served on the Mathematical Modeling sub-committee of the
General Education committee, and I wrote to the chair of statistics and
said, "Look, there's going to be a new general education requirement for
mathematical modeling. You might consider proposing a course for it."
He wrote back to me, copied to the chair of the Math Department, thank-
ing me for the suggestion. The then chair of the Math Department called
me into his office and said, "If you're going to encourage statistics to offer
a mathematical modeling course, you should leave the Math Department
and go to statistics."

Now the then chair of the Math Department was actually very supportive of me. There wasn't a big problem. He just thought that I was a naive

new professor in the department, relatively new, and that I did not understand that math owning the mathematical modeling requirement was essential for the Mathematics Department's wellbeing. So, then, after this, I did two things. One, I looked up the mathematics requirement at every other large public state university and found that no one had a requirement as intense as our mathematical modeling requirement. The other thing I did was I make myself chair of the mathematical modeling sub-committee. I shepherded some things through, and I failed to shepherd other things. Since then, I've come to realize that the chair of math was wrong. In fact, it probably would've been better if we did not own math modeling even then, because it would've changed the dynamics over the coming years and may have prevented the situation that we are in currently. But at any rate, math lives on the School of Business requiring finite and brief calculus, and we will continue to live on the School of Business but more and more other students are finding ways around the mathematical modeling requirement, like taking finite at Ivy Tech Community College and transferring it in.

RH: My journey started with understanding engineering education after I finished my undergraduate education. I first went into industry, and I was very excited to be an engineer. I thought I'd start making things and solving problems. I graduated after four years from one of the premier institutions in the world. In fact, the Microsoft CEO an alumnus of our college. That's an indication of how well known my college is and its reputation around the world. Still, our education system is very traditional, and there was no focus on developing my skills to be a working engineer.

So I graduated, and I got a job. My new company trained me for three months on what to do. I did what they asked me to do. At that point I realized that my four years of education had little do with the work I was actually doing. That got me thinking. Imagine a country like India, where there are more than 4,000 engineering institutions and so many students graduating every year. It's such a loss of effort, such a loss of opportunity, since the institutions themselves aren't actually nurturing students to become good engineers. That led to me joining a student activist group right after college, where I worked with them for two years thinking about the state of engineering education and the kind of solutions that students could affect as change makers.

While I was working with the student organization, I got to know about the program at Purdue, and I applied to their PhD program. My intention was always to return to India and try to impact change. Whatever I'm doing with my institution, I will constantly think how this change can be scaled and what challenges we need to address. As I make change, I want to make sure that the change sticks at my institution. I want to learn what's working and what's not working because there's so many other institutions that need the support. And I'm uniquely placed because I have the background in engineering education now. When I was at Purdue, every grad student had different reasons for why they joined the program. My reason seemed very different from those of other people. I was asked, "Really, you think you can do so much with this degree back in India?" And I was like, "I think so. I think there's lot to be explored, but I definitely think that has to be the focus."

So the idea of change had already always been there. And I think that's what I'm going to stick to for the rest of my life because there's so much to do and the intention of doing the PhD was really to impact large scale change. But I don't think I'm alone. I think I'm just a facilitator, a change maker. There are many more change makers around. I believe it's an opportunity for learning. And I think the most important component is finding the right people, talking to more people, maybe getting into a policy space where a lot can happen. Currently, I'm still at a place where I'm learning as much as possible. But while I'm learning, I constantly keep thinking, "Can this be scaled to different systems?" When I say I can "scale" it somewhere else, every context, again, in an organization is different. I may have come up with some model that works in one place, but can it be translated to another institution? And what kind of change barriers can we anticipate there? Because those will be key when you want to transfer the learnings to different settings. Currently I focus my efforts in two areas: operations and human resource development. With respect to operations, the entire goal is to implement the systems and processes in place that are in line with the necessary requirements of the accreditation standards. India has two accreditation bodies. One is the National Assessment and Accreditation Council, the Indian government's initiative that accredits institutions. NAAC gives different grades on the quality mechanisms. Then there's the National Board of Accreditation. This is particularly for engineering, and they accredit programs, rather than institutions, much like ABET does.

In the last few years, all institutions have been aware that they need to be accredited. If the institution is not accredited, then they understand that the regulatory bodies are saying that they aren't good enough, and they need to improve their quality standards. The National Education Policy also recommends that all colleges become accredited as soon as possible. Now parents are aware that they should send their students to an accredited college since this is a way to ensure the quality of the education each student receives. Unfortunately, there is a persistent belief that while the accrediting agencies should be ensuring quality improvement, they have often been guilty of focusing solely on assigning accreditation without driving quality improvement.

You can see how this would affect how an institution approaches their work on accreditation. If the focus is on just getting a stamp of accreditation, then I might apply for accreditation, work for two or three months, and create a mountain of documentation, sometimes ethically, sometimes not. For example, even at my institution, one week before the accreditation agency was supposed to arrive for the site visit, some faculty said, "Oh, we have to put a sign for fire extinguishers all around the campus. And all the students have to be trained to use it." I pointed out, "We have the extinguishers, but we haven't trained the students." So we did it immediately, but it was a low level priority.

Now, my intention was to communicate to my team that we needed to engage with the accreditation cycle not just for the sake of getting accreditation, but for the purpose of improving educational quality at our institution. These are the quality benchmarks they've set because if you do this intentionally, holistically, then your quality will improve. Accreditation is just a step in the process. It's not the final goal. A big difference between the US and the Indian accreditation processes is that the US institutions usually have support staff who can provide help with the preparation of accreditation documentation. In India, the faculty have to do all that because we don't have the financial resources to have support staff. So now, when I ask faculty to do something related to accreditation, I ask them to document it properly, through a quality assurance system, which audits every department three times every semester. With these audits, I'm trying to put in measures to make sure we remind people that these documentations are missing or these kind of things are what you've not done and we are expecting you to do these things. Now, even the audits inside my institution have become in such a way that, oh, before the audit, one day before there's a whole chaos in the department saying, "Oh, audits are coming so we have to do everything." At the top tier universities in India, faculty, staff, and administrators are self-motivated and do things in their own time, but they also have resources that they require for support staff, et cetera.

So this was a big change problem, because how do you change this? Everyone is so entrenched and used to this mindset. About 90% of the institutions in India do this. So what I'm asking them to do, to plan ahead, is something which they've not even thought about, and they don't even think it's needed. Faculty look around at other colleges who are not doing it, some with better reputations than ours, and they see that the other schools have been doing well. The faculty ask, "Why don't we do the same thing?" So this has been a major challenge. And for that, we had to really ask everyone to rethink fundamentally. As a result, I need to communicate a lot.

Right now, some faculty and staff say, "Let's do some certification programs because NBA has asked us to do them." The communication is itself a problem. I emphasize that we aren't doing this for NBA; we are doing this because it helps our students. If my institution does certificate programs that can help students build better skills that might be required for industry, and it'll get them better prepared, it'll lead to better jobs, and students will actually graduate with something that they really like. And that's such a big culture change that has to happen. That in itself in my institution is a huge task. Now imagine kind of seeing how this can be translated to different places. The Indian educational system that the British put in place was created mainly to ensure that Indian people followed orders, which is why in the system currently, we administrators give orders, and faculty follow them. They expect that the boss will give them the orders, and they will do it. And many of the bosses like giving orders, because it puts them in a position of power that they enjoy. That's why it's such a mammoth challenge. It takes all my efforts to constantly make my people think. In fact recently, we're working on a series of visioning exercises for our area. We've defined our organizational vision and mission, and we've also worked on our individual personal goals. We've come up with a strategy plan that involves every faculty and staff member. They are involved in developing what the strategy has to be. And once the strategy plan has been built for each of the key objectives, we brought in the relevant stakeholders. We came up with a plan on what should be our measurement indicators and what should be our targets. This is now the direction we will follow.

Now, by doing this, how does it really transform you at a personal level in your personal lives? What happens and how can we kind of really align that? And I tell my people, if it's not aligned, let us know, because it has to be aligned. If it's not aligned, then it's a leadership problem. It's our job to make it aligned because when those two are aligned, then you will feel self motivated to do your work because you're doing it for yourself. This is, again, easy to say, but we've recently finished this exercise and now we'll have to communicate it again and again. I hope that my faculty might start getting this, but also my leadership team has to get this too.

TW: As part of the merging of departments, there were specific challenges related to the fact that we didn't have micro electronics, so we didn't actually have that physical dimension with semiconductor fabrication facility. At ERAU, part of the circle is actually missing, so you have to visualize it. I would try to use it to appeal to students again, because if they don't come in, they don't really know what they want to do. And so, we're trying to give them these concepts, this broad area. And it evolves into things like managing software and systems, which again, involve these conceptualizations of what you're doing and how these concepts and modules interact.

I don't know if it really took or not, because it's a hard sell. You've got just some way to keep them thinking they're part of the same team. And that really brings up to me the bigger issue, which is that faculty by and large don't know how to work in teams. We are allegedly training students to work in teams and yet, faculty haven't worked much on teams, either in industry or in academia. I believe you can't share knowledge of how to do things without having the practical experience of having done it. You can guide that and grow into it. We've all done it for different topics. It's definitely not part of the reward structure. Faculty rewards are built around individual performance and how many journal articles you have. Whether journals are a really big thing in your field or not, it doesn't matter, does it? It's what matters to the promotion and tenure committee.

When our department received our NSF Revolutionizing Engineering and Computer Science Departments (RED) grant in 2019, we tried to implement scrum, which required the faculty to work in teams. And immediately you find out, well, faculty don't know how to work in teams. As I learned after we're doing it for a while, scrum really ought to have its own architecture. It has a very much of a command structure where this product owner really is the person who gets to make the decisions about what we're going to work on this next sprint. The team gets to decide how to do it. But the project owner gets to decide what the priorities are.

You can't do that in academia. There's no command structure in academia. You can't tell people what to do. You can tell them you're going to give them a terrible class at a bad time of day if they don't act right. But you're not going to do that because you don't want to punish the students. The Dean is always saying, "Well, you control their schedule, and you have to put them in what they can do."

Good grief. You can't jerk them around. You can't say, "I'm going to punish you by making you teach freshmen." That's not punishment. That ought to be a gift. My gosh. I had never taught freshmen, and I was scared going to EGR 101, because again, I had taught grad students and upper division classes and stuff like that, but really, it's the best thing in the world.

With regard to faculty operations and the RED project, I should have spent a year or two just getting the lead team working like a team. It needs to be a model, and we weren't a model, we weren't. It's going to be hard because, as you know, faculty believe they know what's right. As a change maker, you have to acknowledge that, and you have to hear what they think. But can you put into place for faculty a command structure that says, "This person has the decision authority, and you will do what they tell you to."? I don't think you can. Can you make a faculty that's agile and does experiments and learns things? I think that's a much more, a greater opportunity there to bring broader concepts of agile into the academic community and also to teach students. Again, it's akin to thinking critically. It's a way of slowing down and making sure you're focusing on important things and learning things. That you're trying things and learning things from what you're trying, not just randomly doing things explicitly.

And we'll try this for this. See what happens. Observe. Make some evaluation. Make your next decisions using this new information. So it's probably more like a lean startup than implementing agile strategies. There were some early signs that the faculty who had been funded to work on lean startup or agile teams weren't prepared to work as a team. People would dial into the meeting and never speak up. There's always that, just freeloading. No engagement. Just not doing the training. And then, if they got engaged, they'd want to run it all. They'd want to be in charge. They're not going to do it, unless they're going to be in charge somehow. We had this thing teed up, finally ready to go into teams for the whole faculty, and that's when I got fed up and said, "No, I can't, I can't. These people are not playing right by my values anymore. We have a disconnect." I don't know if you and I have talked about it. I went in one Friday and they were looking to find somebody to teach a class in-person that had been developed online and taught online by a recently tenured faculty in the engineering fundamentals department, who wanted to teach online again, because his wife is immune-compromised. But the administration said that this faculty member had to teach face-to-face.

And then I had other colleagues who had immune-compromised family members, but we weren't allowed by the administration to even ask students to have masks on in office hours and crap like that. And it's just at some point, it's, "No, I can't." The guy they wouldn't let teach online, even though he had developed a class successfully, and it was a class to help ensure that students at risk succeed. He's invested in these students. He will not walk away from these students. And I'm just like, "Dude, aren't you looking for a job somewhere else? What's going on here?" It's the students. It's this sucker punch that we all get invested in the students and then they levered that, obviously.

JW: **How does your disciplinary training influence your approach to change?**

SC: One of the things that Carl Weiman who put together the SEI was really focused on was having data that a scientist would not refute. He would often argue about the data but I think that that's actually wrongheaded because physicists argue about the data, not because they disbelieve the data, but because that's what they can hook on to argue against active learning. They're using a rational argument to justify what's not necessarily a rational response. So, arguing it at the rational level never quite works and that's why I think hearing the student voices is always really powerful. It speaks to our reasons and beliefs about teaching.

We often would see that, once a professor started using clickers, which was a common thing that we did at CU Boulder. At that moment, the faculty member would realize that while they thought that the students were achieving gains, they weren't actually. Once you just get that little window into students' thinking, the faculty member thinks, "Oh, I was fooling myself that they were getting it." And so, a lot of these formative assessments do give you that window that wasn't there in the non-active classroom. All of that to say that there are other mechanisms for seeing whether or not things are working in an active classroom that aren't there in the traditional classroom.

There's also a lot written in the SEI Handbook about challenges and the strategies to overcome those challenges so that's another place to look: challenges for the teaching and learning fellow, the post doc, the faculty in the department, and the people leading the change, challenges at all those levels. I'd like to talk a bit about the experience of the postdocs in the SEI model. We didn't just grab a person and plunk them into the department with the assumption that they go off and do good things. They're an individual with career aspirations and challenges, and they're being put into this difficult spot of going to work in a position that doesn't actually exist in the department. So, they definitely had challenges, often around imposter syndrome, asking questions like, "Do I know enough and how do I develop credibility and prestige within the department?" So we had suggestions for them. We would suggest that they'd give a talk on their research, non-education research, because most of them were coming from more traditional postdocs. We would suggest that they talk about their traditional research and run some faculty meetings around education work just start to establishing connections with people in the department. You may not know this, but I was a Peace Corps volunteer in Africa and it always really struck me as very similar to being a Peace Corps volunteer. You get helicoptered into this village and go figure out how it works and find out what they need and make good change happen. So, the first year of Peace Corps is figuring out who everyone is and what they need and then the second year is when you actually start to do work. I felt like it was the same with these postdocs. Go slow to go fast, take some time to figure things out. One thing that was challenging, though, is that we didn't want the SEI money to be spent on hiring teachers. We didn't want the postdoc to be just dropped into a course and teaching, but then we found that they actually needed that teaching experience to gain credibility. And some of them just hadn't taught very much or at all before so there was a fine line there, giving a little bit of teaching experience, but also ensuring that there was supervision of that postdoc.

We made sure that they had a department liaison, a person within the department who was in charge of helping to negotiate which courses are being transformed, which faculty are going to work with them to start so that they didn't have to negotiate that themselves at the beginning. And then, over time, they cultivated relationships, so eventually who would work with them would become more organic and emergent. But I think that that idea of the department liaison or department local leader is something that's important for change initiatives in general because, otherwise, you get this diffusion of responsibility.

Especially in these cross-STEM departments, having that person who knows the department, has some leadership chops, maybe not the chair, but someone with some expertise and well regarded by the faculty. When that liaison wasn't somebody who was organized and well regarded, things often would fall apart. Throughout the project, the main thing we wanted to know was, were the postdocs who were hired by SEI employable in their disciplinary field, or were we breaking these people's future career aspirations? And because it was completely unknown at that point (roughly 20 years ago when the program started), we wanted to be sure that these people were going to be able to use this experience to become employable. The answer to that question was a clear "yes." We have collected what kinds of things those people went on to do. We found they are directing teaching and learning centers, serving as education research faculty, they're teaching. There's just a wide variety of careers that somebody with expertise in the discipline and in teaching and learning can pursue. They are very hirable, which I'm sure doesn't surprise you.

I'm a bit of an odd duck in the things that I do but I'm not the only one. There are lots of different career paths. Early in the initiative, when things weren't as well organized in terms of supervision and helping them navigate difficulties, a lot of postdocs either quit or threatened to quit. There was a lot of unhappiness and stress early on. So, we learned the hard way, and this was concurrent with me actually being a postdoc as we were figuring this out. But we learned the hard way that, if you don't attend to the care and feeding of your postdocs, then it just doesn't work, it's a very unhappy situation for them.

One of the projects I work on nationally is the workshop for new physics and astronomy faculty. And so, we're really devoted to the disciplinary professional development to show people how to teach within the discipline. But a lot of people say they just don't get as much help from their teaching and learning center, it's just too general. I feel like that disciplinary professional development, be it through the postdoc, be it through disciplinary workshop, prepares a teacher to understand the language of the teaching and learning center. I think it's a transfer problem, right? The first time you learn something, it needs to be near transfer, transfer from a situation you're familiar with, the teaching of physics or the teaching of biology. And then, when you go to the more general workshop, that's far transfer. They're giving you more general information that you then have to transfer further into your discipline. And once you're more an expert, you're able to do that far transfer. It's how people gain expertise, and teachers aren't any different. I prefer to call it transfer because that's the language of teaching and learning. Just the idea of transfer has so much packed into it that's really valuable in understanding how you take on expertise. Julia, you and I met through the Accelerating Systemic Change Network (ASCN), which has been actively pursuing academic change across STEM education. What I like about such cross-disciplinary efforts is that it brings together so many diverse fields. It's clear that different communities are struggling with the same issues, and here's a place where they can all get together. Do you know the idea of the small world networks? In *Small Worlds: The Dynamics of Networks between Order and Randomness (Princeton Studies in Complexity, 36)*, Duncan Watts describes communities of people as "A knows B and B knows C and so A is more likely to know C." Academic change has been like that with these clustered communities, and I feel like ASCN is an opportunity to start to make bridges and connections between those clustered communities so that we become, ourselves, a small world network of change leaders. I feel like that started to happen. My involvement with ASCN has made it so that when people come to me when they're trying to find someone who knows about X, I am able to send them to people that are outside of our community, like you, right? I would not have known you except for ASCN and so now I have this bridge to you and your work. Those little connections are so helpful in starting to broaden the network. Because now, because I know you, I can be like, "Well, do you know someone else who knows about X?" and you can connect me to a whole new set.

EH: I think that I do well on committees, that I have good ideas, that I listen well and synthesize other people's ideas well. I can tell you other reasons why I do that. I can call votes before they're taken. I was on the College Tenure Committee, and there was a difficult case. I didn't realize that the votes weren't public. So, before the vote was taken, I said, "We're split 50–50." We were split 50–50. When I was elected chair of the Math Department, I went into the mail room and I went through every tenure track faculty mailbox, and I said to myself how they were going to vote.

Maybe I didn't have them right by name, but my numbers were dead on. The Math Department constitution has rules about how things happen depending on what the vote is. It happened exactly on the level where I had counted the vote. I can't do that, of course, across campus, on major issues, because I don't have that kind of personal connection. But in the Math Department, when there's about 40 people, I pretty much know them, and I know what they think. I don't think most of my colleagues know what most of my other colleagues think. But I do. This may be an interesting question about whether it's a feature of my autism. So, there's this assumption that autistic people aren't empathetic. Then there's another idea that they're super empathetic. I don't know, but it's some-thing that I have been able to do in small enough groups.

Before I was elected as the chair of the math department, I wrote a to-do list, a SWOT and a to-do list of what I wanted to accomplish as chair. Oh, I ticked them all off pretty much. In fact, I ticked them off up so fast even before I came up for re-election as chair. Math has a very lovely system for reelecting a chair. During the beginning of your third year, they hold two votes. The vote of the faculty is whether the chair wants to con-tinue. The vote of the chair is whether the chair is willing to continue. And a three-person elderly men committee, except the one time an assistant professor got on it, opens both sets of votes, and only announces whether there will be a chair election or not. They announce nothing else. So, they don't say whether it's because the faculty don't want the chair to continue, or whether it's because chair doesn't want to continue.

But before that vote, I had to ask myself, "Is there anything else I want to accomplish as chair?" Because I've pretty much done it all. And then I came up with a few more things to do. So, I wasn't blamed up for the last two and a half years. But I did make myself a list of things to accomplish. They weren't dramatic things. So, that's the other thing is that lots of peo-ple want to go in. In fact, it's quite clear that our current provost needs to effect dramatic change. It is my experience from talking to head hunters for dean positions that I have to have implemented a dramatic, huge change in something in order to be considered dean material. I hope that I made no big change. What I did, 1,000,001 little tiny things that made things better, 1,000,001. Just a vast list of tiny things that made the world better.

RH: Our office in the university has around 150 people. And my leadership team is around 25 people. In order to get everyone moving forward with the strategy, you have to make a choice. You can push them through fear, which is never sustainable, or you can inspire them with trust, which is sustainable, but it takes time to build. So if you kind of build trust with each one of your people, and they believe that if you're asking them to do something, it's going to lead to also their personal and professional growth, then they definitely will align with whatever you want to do. So it's that relationship that you have to build. You also try to empower them and make sure that they feel respected, they feel loved, and they feel acknowledged so that they start believing in you. That's how it works, by bringing in behavioral change.

I think the first thing was setting the expectations straight away, what are your expectations and keeping people self-motivated. I want to encourage everyone to reflect on their performance throughout the year and assess for themselves whether they are doing a good job or not. I actually picked up this tool from Deloitte. I worked there for a year after graduation. And we had a very comprehensive HR system where at the start of the year, we had to meet our mentor, and we had a goal setting process. We set our goals for the year. Then we would have a midyear review. And at the end of the year, we checked off if we'd met our goals by a process of self-appraisal. So I've been implementing something similar at my institution. The first step was to set expectations. We told them very clearly that as a faculty, there are three things that they are expected to do: teaching, research and administrative responsibilities, each with different parameters. On the appraisal form, each parameter is given different weight. Each faculty member sets their goals for parameters like teaching effectiveness, innovative teaching and learning, and so on. It's through this form that we can set clear expectations and communicate their importance. We communicated early on that this is what each faculty member would be measured on. The form functions like a rubric so they know what is expected. Then it's up to the individual.

At the start of the year, faculty members set their goals, and their deans review them. And then in the mid-semester, we have reviews just to remind them about what they are working on. At the end of the year, if they meet their goals, then they're rated as good. If they exceed their goals, they're rated excellent. And if they kind of don't do it, then they could be assessed as average, bad or poor. So these are the ways for us to empower them and let them know these are the expectations. They don't have to constantly wondering, is anyone watching me? Should I be doing the work? No. This is what is expected. They know that if they do their work, people will recognize it. With the systems approach, a faculty member has everything in place. They just need to follow their plan. And if they need resources or help, they can check in with their team. The team is also checking in regularly so that it's not at the end of the year, they're like, "Oh, what the heck were you doing?" So throughout the year, we have reviews where we remind them and guide them on what they're doing well, what they're not doing well. And at the end of the year, they just have an appraisal and it's always systematic, all numbers driven, very less qualitative. That is what we want to make it because qualitative brings in feelings and biases in people, and we didn't want that to happen.

So this was something I started implementing, and it took a while. It took a couple of years for the process to also just kind of systematically come in place. And now we are coming to a point where people are starting to think, "Oh, these are the goals I set at the start of the year. I have to achieve them." I should add also that the expectations for the role of a teacher in India changed significantly in the last 10 years. Prior to this, the individuals who went into teaching were the ones who could not go anywhere else. For example, someone who couldn't get a job in industry would become a teacher, thinking that since they were taught engineering subjects, they could replicate the same information for their students. Now, the demands of being a teacher have changed because you have to also do research. Research is of utmost importance now. There are also many more administrative responsibilities with documentation, accreditation agencies, etc. So the responsibility has changed.

We always work through people. It's not that I'm building a product and I'm putting it out there. It's the people that matter. And that is what I want to share, that it's the people that matter. So we have to value them. We have to appreciate them, and we also have to recognize that they're all different in different ways. So it's very important to kind of build that trust, communicate effectively and be in a position as a leader to empower them. That is what is going to lead to true change because change always has to come from within. It can't be prescribed because prescribed change is not really change. Using prescribed change, you're putting faculty into a forced setting, which they might or might not even want to be in. So change has to come from within and that you'll have to empower them.

Change takes time. It requires you to be very patient. The question is, how long will you be in this organization? If people come in and then maybe in a couple of years, just leave to another opportunity, they really can't be motivated about change because the moment they get started with initiating something around change, by the time they want to leave, they're not even maybe scratching the surface. So change takes time, which is why there needs to be leadership who stays, at least four to five years. You need that timeframe at least to initiate something, observe it, and sustain it. And that I feel is another thing, which is very hard about change and which is why people don't really want to touch upon it much because that's the harder way to do things. And why it is the hard way? Because it's harder but also the more sustainable way for the organization.

When you're significantly changing behaviors in an institution, you need to highlight the new behaviors and acknowledge them. So we have something called monthly spotlight awards where we draw attention to our priorities, like a star innovative teacher or a star researcher. Anyone who's taking ownership about something gets recognized. Every month, we are recognizing people and acknowledging behaviors that are standards, and we want everyone to kind of keep engaging in them.

As a result, we realized how important recognition is for people, and how much it matters to them. It's not even that we give them any money and stuff. It's just a small plaque, a small award. It just makes hell lot of difference for them. And that was, again, an interesting thing for culture building because these are cultural attributes for the organization. And I felt that was very important to build a culture based on what we needed to change. My job has definitely been very challenging, but also very exciting because I'm questioning almost 25, 30 years of training. And sometimes people are almost 45 years old and they have worked in the system for so long. It's challenging, but I know that if I figure it out here, then I have a lot of answers for the other context. So that kind of keeps me going.

TW: I've got lots of advice for lots of people! Seriously, I saw this model when I was a graduate student. And so again, I'm using what I saw when I was in grad school as a model for how to do this now. But this worked, and this was from the guys who put together the first five year master's at MIT, in which students would get both a co-op experience and a master's degree. They put together this unified program, and I watched them do it, two of the guys on my committee, Bill Secret and Cam Sheryl.

I thought their strategy was great. They had lunch with almost every faculty member in the department over a course of about a year, pushing this idea. Talking it up. Getting buy-in, and it wasn't until they had buy in close to a super majority that they knew that it was going to be a selling idea. But then they go to the Department Chair, who was Paul Penfield at that time and say, "Hey, this is written. The faculty will support it," and then Penfield puts it through the formal process.

So, one thing I think we do as change agents in academia is we latch onto the formalities of change, like the formal proposal document. My advice is, don't even write the proposal until the change is ready to go. Everybody ought to know what's going to happen. What the change is. Nobody wants to read about the change through the proposal. You have to brief people about the change through informal, non-structured ways. Find ways that they think they can improve it. Ask for their advice. All those classic things about how to get buy-in from people. That's what is valuable.

And I wish I had had time. I wish I had done a better job of that on the RED Project. We just threw our proposal together from a previous one over a summer, but there wasn't time for schmoozing and getting buy-in and all the real work that has to be done to get your colleagues to be comfortable. They need to be. It's unfair really to ask them to just to trust you. They are smart people. There's a reason they think they know everything because they're smart. And you have to work with that.

In contrast, when we put together the doctoral degree in our department, we did that. We had buy-in. We talked to everybody. We took our time. A bunch of people worked on a proposal and then a bunch other people worked on revising. And so it really got a bunch of buy in and stuff, but the RED project was rushed. And even within the proposal team, I wish we had worked harder to build a shared sense, the shared vision. That's key in successful agile development, to have a shared vision. I think that's it.

I always come back to the ending of the novel *Lonesome Dove* when the main character says, "It was a hell of a vision." It's always great being a visionary because you see all these possibilities. Yes, you see all these possibilities, but you just can't reach all of them. It ain't possible.

2 Getting Ready for Change

INTRODUCTION

In this chapter, we'll focus on the earliest stage of academic change, before you even initiate a change project. By focusing on preparation before you start, you can better prepare yourself for the work that lies ahead. The tools you will use in this stage are:

- Understanding yourself as a change maker
- Assessing opportunities for change on your campus
- Evaluating your readiness (and your team's readiness) for change
- Scanning your campus environment for potential challenges to your change project
- Planning a roadmap for your change efforts

These tools apply to whatever change project you are embarking on and whatever level you believe you are as a change maker. Some readers may be considering a change project and haven't yet made a start. Others may be in the early stages of a change project. Still, others will have tried and failed, or tried and wish to make their change work easier next time around. No matter what change you want to make or your level of experience with academic change, the tools outlined in this chapter will be useful.

You'll be asked to practice each tool by responding to exercises. Your responses will be different from a colleague, either in your home department or on a different campus entirely, and that is exactly how it should be. The purpose of this chapter is not to present a formula that worked for one academic change project in a single context. These tools will be most useful for you if you learn them, then adapt them to your specific circumstances and context. They become your tools when you customize them for your purposes.

As indicated in Chapter 1, each topic addressed within this chapter will include an opportunity to reflect and think about yourself and your change work. The reflection might, at first, seem unexpected, perhaps not even relevant. For example, it may not seem germane to begin an exploration of yourself as a change maker by asking you to write a story about yourself in the style of a fairy tale. Be assured, however, you are being asked to reflect in order to better prepare you for the work that lies ahead. These reflections are relevant and should stimulate your creative thinking about your academic change project. In the process, you will be developing habits of mind that can help you understand and remedy the challenges that stand in the way of reaching your goal. It might feel uncomfortable at first, but, as is the case with learning any new skill, your practice will ensure your progress.

DOI: 10.1201/9781003349037-2

FIGURE 2.1   Getting ready for change. (Used with the permission of the author.)

Finally, your work with the change tools will be reinforced by the Change Maker Interview that appears at the end of the chapter. You will meet Dr. Jeremi London, a working change maker who has great insights about getting started with change. As you read the interview, you will see how the change tools from the chapter have been applied in real-world change projects that Jeremi has worked on.

This chapter begins with the image of the change maker opening a door onto new possibilities for change (Figure 2.1). Now it's your turn to open the door and walk through.

YOUR CHANGE AUTOBIOGRAPHY

Summary: Using a fairy tale template, you'll write about your experience with academic change in order to empower your development as a change maker.

> Fairy tales are more than true—not because they tell us dragons exist, but because they tell us dragons can be beaten.
>
> —Neil Gaiman

Your first step in getting ready for change is to take a moment to look back and reflect on your previous experiences with change, by writing a change autobiography. Your previous experience might be at a small scale. It could be extensive. You may have never ventured into a change project at all. Whatever your experience, your change autobiography can help define what you did in the past and what worked (or didn't). Your change autobiography documents any experience you have had with change, so you don't need to limit yourself to experience within an academic context. The goal is to bring out what you attempted, what worked, and definitely what didn't. As a result, you can cultivate greater self-knowledge about what your strengths as a change maker might be and what areas you need to work on.

And rather than have you write your change autobiography as a realistic narrative—first you did this, and then that, etc.—you should write your change autobiography in the mode of a fairy tale. The fairy tale style is quite effective for this task. By keeping your focus on the fanciful, you can strip away everything from that experience except the essential elements. You can concentrate on what you can learn from it. When attendees at my change workshops are asked to begin their fairy tales, some individuals resist. How does writing a fairy tale help them with their change efforts, they sometimes ask. In such situations, it is important to emphasize the ways in which such writing strips away every irrelevant consideration and helps the writer focus. Perhaps you tried to implement a new pedagogy with colleagues in your department, only to be challenged by a more senior professor. Consider how this experience is transformed if you write about it as a version of Jack and the Beanstalk. Or you proposed a change project that required the support of your dean. Your dean expressed support for your ideas when you met with them in their office but expressed doubts and hesitancy in a more public setting, as if they were wearing grandmother's clothes to hide their wolfish side.

Your change autobiography story is really a story about yourself as a change maker, where you started in your path, what you learned along the way, how you applied those lessons to each subsequent change project, and how that knowledge builds from both the mistakes and the successes. In addition, when you tell a story about yourself as a change maker, you are also telling a story about your change project as well, and such project stories are an important tool that you will be introduced to later on, in Chapter 4. For now, the focus is on you.

Before you get started, take ten minutes to brainstorm on possible change stories from your experience that could serve as the basis of your autobiography. Remember that you can use any change experience, everything from deciding to

change your regular coffee shop to planning a major revision of the general education curriculum.

In the space below, consider a time that you made a change:

What did you do?

What did others do?

PUT THE TOOL TO WORK

With the brainstorming to get your started, your task for this section is to write your
change maker autobiography. You can choose any previous experience you have had
with academic change making.

- Maybe you'll focus on the semester when you developed a problem-based
  learning approach for the required course for non-majors.
- Perhaps you'll revisit the time you chaired the college committee on insti-
  tuting a particular change that was intended to improve recruitment and
  retention of faculty.
- The story you choose is up to you.

As you write, consider what you have learned about yourself as a change maker, in
either the professional or the personal realm:

- What are your strengths as a change maker?
- What are your weaknesses?
- What do others see in you that makes you, or could make you, an effective
  change maker?
- What skills do you bring to change work? What expertise do you possess
  that would be useful in a change project?
- Was this change successful? Why or why not?

You may find it beneficial to consider a standard fairy tale as the starting point. Your
version may have a "just right" theme, or a theme of finding the path through the forest, avoiding trolls and other threats. No matter your choice, you can use the story as
part documentary and part wish fulfillment. If you are less comfortable in the realm
of fairy tale, feel free to adopt a different style. You may prefer the science fiction
genre, so follow your interests as you write your change story set on the Starship
Enterprise or on the planet of Tatooine. You can also approach your autobiography
straight on as a record of the events that transpired when you attempted to make
change in your department or on your campus. The choice is yours!

READ MORE ABOUT IT

If you'd like to read more about the power of fairy tales, check out Bruno Bettelheim's landmark book, *The Uses of Enchantment: The Meaning and Importance of Fairy Tales* (2010). You'll never read Little Red Riding Hood the same way again!

Bettelheim, B. 2010. *The Uses of Enchantment: The Meaning and Importance of Fairy Tales.* New York: Vintage.

EMERGING OPPORTUNITIES FOR CHANGE

Summary: In this section, you'll use the Emerging Opportunities tool to identify and assess opportunities for change that currently exist on your campus.

In order to get ready for change, you've reviewed your experiences with change and what you have learned from these experiences. As you continue through this book, you will have additional opportunities to build your skills in self-reflection, all with the purpose of expanding your change toolkit.

Reflecting on your change-making self is one component of change readiness. Another component is recognizing the opportunities for change that exist all around you on your campus. You may have opened this book with a clear change effort in mind. In that case, you may not feel that you need any help in identifying opportunities. You may be planning your first change project, but it won't be your last, so opportunity recognition is a useful tool to have in your toolkit for now and for later.

We'll start development of opportunity recognition skills by dipping into a field that you may be unfamiliar with: the field of entrepreneurship. If that word raises your hackles, be assured that the purpose of this chapter is not to start you in a business venture, unless of course that is the change you have in mind. The tools of opportunity recognition are often addressed in the study of entrepreneurship, but similar tools are also a part of many other fields as well, including your work as a change maker. For individuals and teams planning an entrepreneurial venture, the Business Model Canvas (Figure 2.2) provides a useful, visual template for the enterprise they have in mind. First proposed by Alexander Osterwalder in 2004 and later expanded in his book *Business Model Generation* (2010), the canvas identifies nine elements that comprise the foundation for a successful business model.

The inspiration for the canvas comes from the tendency, Osterwalder says, for entrepreneurs to be "held back in our thinking by status quo. The status quo stifles imagination." In contrast, the canvas encourages its users to consider the necessary elements that go into a new venture, everything from business activities to potential partners.

Now, before you object to the business model canvas, saying perhaps, "I'm not interested in starting a business … I am trying to change how my department teaches Differential Equations!" consider a variation on the business canvas idea: the Social Enterprise Canvas.

Like the business model canvas, the Social Enterprise Canvas is designed to help individuals and teams define elements that form the foundation of their social, rather than business, oriented enterprise, but the list of elements is expanded from 9 to 11, a sign that social ventures require additional considerations. So, if you are interested in, for example, establishing a community-based recycling program or other operation that will be not-for-profit, then you can complete the canvas, stimulate your creative thinking, and avoid pitfalls before you even label bins for aluminum cans, clear glass, or newspaper.

These canvases also can inspire for good thinking about change projects. And for that reason, it has been adapted for the purpose of stimulating your thinking about what needs changing on your campus, into the Emerging Opportunities tool. If you already have your change project in mind, you'll be using the canvas to inspire deep,

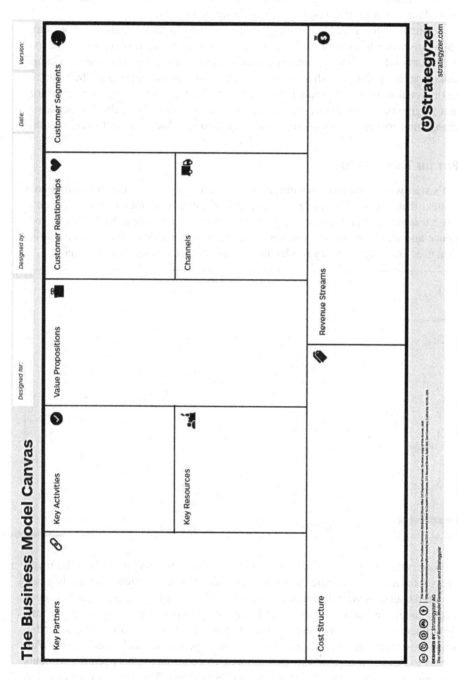

FIGURE 2.2 The business model canvas. (Used by permission from Strategyzer in the Creative Commons.)

clear thinking on elements of your project that you haven't yet probed. Even if, after you use the tool, you don't see much alteration in your plan for your project, you may find additional areas that you can address in future projects.

Seeing the Emerging Opportunities tool in action is truly inspirational. For example, in my workshops with graduate students, they have used this version of the tool at the start of their academic careers. Many of them would not call themselves change makers at this point, but when given the chance to consider aspects of their current institutional setting that irritate them, they had plenty to say. It may even be important to give them the chance to use the tool and envision themselves as academic change makers before they are hired into their first job. They are our future, after all!

PUT THE TOOL TO WORK

Let's start with some brainstorming. List five elements of your current institutional context that "irritate" you. For example, you might cite a problem you experience when teaching a required course in your department, or a problem that students experience and they complain to you about. If the term "irritation" seems too negative, then use "challenge" and try to identify challenges that you see in your context.

1.

2.

3.

4.

5.

Next, select one of the items on the list and assess its direct impacts on both you and on others. For example, you could consider that the content you teach in the required course in your department depends on the quality of the prerequisite course that is taught in another department, but there is no coordination between the two departments. As a result, there are serious gaps in students' skills and knowledge when they take your course, thus impacting how you must teach your course. With regard to others, you could cite that the lack of coordination between departments has caused tension among the faculties and for students. This tension is reflected in poor course evaluations in your departmental required course.

Impacts on Me	Impacts on Others

With respect to this item, now note anything relevant in the following categories: activities, relationships, resources, and resistances. Using the required course example used above and considering key activities, you could meet informally with a faculty member who teaches the prerequisite course. With regard to key relationships, you could connect with faculty outside of your department who also rely on the prerequisite course. Are they experiencing the same lack of coordination and gaps in student preparation? With regard to resources, you should consider this as meaning more than just money. Reflect on time as a resource, especially when consulting with faculty outside of your department who are offering the prerequisite. How much time will it take for them to make the changes that would benefit both you and them? And what are you willing to contribute to the project in terms of your own time? By considering resistances, you can reflect on the situation of the faculty who offer the prerequisite. How would you feel if someone outside of your department suddenly made demands for change in a service course like this? What additional challenges does the other department face? What has been tried in the past? How did that attempt go?

Activities related to this change—what would
  you do to relieve the challenge/irritation?

Relationships—who could help you relieve the
  challenge/irritation?

Resources—what do you need to relieve the
  challenge/irritation?

Resistances—what roadblocks stand in your way?
  Who benefits from the status quo?

Now it's your turn. Using the item you identified, take some time to assess the Activities, Relationships, Resources, and Resistances as they relate to that item. You might benefit from discussing your analysis with stakeholders who are impacted. Keep your focus on the opportunities afforded by this detailed look at a particular challenge you can see on your campus. Your next step in the Getting Started process is understanding more about your department and your campus in order to align your change work.

READ MORE ABOUT IT

If you are interested in doing more work with the canvas, you can find many applications online. You can find the canvas model on which the emerging opportunities tool is based on The Social Enterprise Canvas, available through Creative Commons. You can also learn more about the application of entrepreneurship principles to change work from Osterwalder and Pigneur.

Osterwalder, A., and Pigneur, Y. 2010. *Business Model Generation: A Handbook for Visionaries, Game Changers, and Challengers*. Hoboken, NJ: John Wiley & Sons.

INSTITUTIONAL LANGUAGE AND THE RHETORIC OF CHANGE

Summary: As a change maker, you need to look carefully at the language the department or campus uses to represent itself. Making use of that language can make your change work more impactful.

You have prepared for your role as a change maker by examining your experience with academic change in the past and by identifying opportunities for change that exist on your campus. The third area of preparation you'll address is communication, since your communication skills will be integral throughout the life of your project. Before you begin, however, you'll consider how your department and campus communicate about themselves. By developing a firm understanding of department and campus communication—both in words and in images—you'll have a useful tool that can help move your change effort forward.

Perhaps you are familiar with this old saying, "By their words you shall know them." It may be a trite phrase, but it holds true: what we say defines who we are. In academic contexts, specific words and phrases form a department's lingua franca, its common language. Each new faculty member undergoes a process of acculturation into their new department and campus. One important part of the process is learning the language and employing it as a marker of belonging (Suddaby and Greenwood 2005).

Consider the following two examples. First, on my campus, students rename difficult courses and identify them by a campus-specific slang. Meatballs=Materials and Balances (chemical engineering). Disco=Discrete and Combinatorial Algebra (math). During first-year student orientation, a new student (let's call her Erin) may not understand these terms, but give her a week, and she will be fluent, using them as a sign that she belongs. Second, specific phrases are often employed to define our campus culture and distinguish it from that of other institutions. The campus community is referred to by some as a "family." For Erin, as well as for some faculty, staff and administrators, the word "family" connotes positive aspects of the campus culture. Like a family, we support community members. We encourage Erin to take on the challenges of pursuing a STEM degree. We prioritize her success. Erin knows that Dr. Martin, her calculus professor, is available to help her with homework and projects. Her Resident Assistant Melissa monitors her wellbeing and shares resources if Erin seems stressed.

This use of language is not unique to my home campus. Similar phrases are employed at campuses that are different in size, student population, and focus. For example, staff in the Center for Teaching and Learning at a large research university told me during a workshop that their campus is "relational" and considers itself a close-knit community. The phrase was reminiscent of my campus' "family" marker but belied the large research university's student enrollment of over 30,000. In both cases, the lingua franca of each institution reflects how they perceive and speak about themselves. And the best barometer of a department or campus' culture is the words and phrases that form its common language.

Of course, the use of the term "family" is potentially quite loaded. How you feel about "family" depends largely on the kind of family you grew up in. For example, you don't fire a member of your family, and yet, college staff, faculty, and administers are sometimes fired. Moreover, the concept of "family" often serves as an effective marketing tool. For example, am institution may target prospective students using the word "family," perhaps to reassure parents that their children will be safe and supported. "Family" may, however, create unrealistic expectations. In order to redeem the promise of "family," for example, faculty may feel compelled to answer student email at any hour of the day or night, upsetting work/life balance within their own families.

In addition to shared language, often-used images serve a similar purpose. For example, we can examine the images used on the college webpage or in mailings to prospective students. At my campus, we are fond of images that depict students building a Formula 1 racecar in the competition teams' workshop, or a faculty member assembling experimental equipment with two or three students. And it doesn't matter if your campus uses similar images. Just like words, images identify your campus to itself and to external audiences.

Both words and images form the language of your campus and can instill a sense of belonging among the campus community members. In order to employ the verbal and visual language in support of your change effort, let's spend some time identifying, then unpacking, the key phrases and images. Linking them to your project signals that you understand the community and how it prefers to see itself. Such alignment also signals that you possess legitimacy to promote the change you seek. Who has legitimacy to speak and to propose change? The one who best understands the campus culture, adopts its language and images, and aligns to current values while demonstrating the clear path from the current state to a revised, re-envisioned future state. Use of shared language and images also helps you create a shared vision for the change you envision, as well as provide you with important text that can be useful in written proposals and oral presentations.

PUT THE TOOL TO WORK

Take a moment and brainstorm a list of words and phrases that you believe are the lingua franca of your department and/or your campus community. In order to determine whether these are the correct words and phrases, consider what you would need to translate if you were hosting a prospective new faculty hire or an external visitor. You can test your choices with a colleague. They may even add a few to your list. Next, consider what these words and phrases indicate about the culture of the department or campus, their positive and negative connotations.

Word or Phrase	Positive Connotation	Negative Connotation
Example: Our campus is like a family	Supportive, warm, open, fun	Relies on hierarchical authority; creates emotional situations; doesn't accept new members readily
Your example:		

Now do the same with one or two images you associate with your college's identity and would be understandable by another member of your community, such as a colleague, a student, or a staff member.

Image	Positive Connotation	Negative Connotation
Example: Students building a Formula 1 car in the competition teams workshop	Learning takes place in a co-curricular activity outside of class; students collaborate	Not all students think that working on cars is fun; image may rely on gender stereotypes
Your example:		

Now let's put the tool to work. By employing the words and images when you are speaking or writing about your project, you signal that you understand how the department or campus works, its generally held values, and your willingness to link what exists now to the project you wish to advance. Consider your change project and how it links to the verbal and visual elements you just analyzed. For the purpose of illustration, I will complete the grid using a hypothetical project I am working on—a proposal to revise a core class in the major with a problem-based learning approach.

Existing word or phrase: Our campus is like a family.	**Alignment with my project:** I propose that we bring problem-based learning into this core class so we can provide students with additional attention designed for their unique needs from caring instructional faculty.	**Put into practice:** Invoke the existing phrase but in my specific, new project context, for example, "I want to introduce PBL to our campus family," "learning PBL is a lot like learning how to ride a bike with the assistance of your parent," etc.
Often used image: Faculty member demonstrates the use of lab testing equipment to a group of students.	**Alignment with my project:** Create an image that shows instructional faculty working with students on a problem-based activity in the lab.	**Put into practice:** Include the revised image in a PowerPoint presentation to the department, adopt the same image for an informational webpage, use it as well in any social media postings you made on the project.
Your examples:		

The examples provided above rely on leveraging positive associations with the words and phrases familiar to your campus context. It is possible, however, to leverage the negative associations for each phrase and visual in order to highlight the specific problem that your change project addresses. For example, your college may have a problem with attracting female students exactly because the Admissions Office relies on the image of male students working on competition cars in its marketing literature. A change project can use that negative example and propose instead that the college use a more neutrally gendered image of an alternative activity. And don't forget how both the positive and negative associations may be useful tools when you begin to formulate a shared vision for change with team members, stakeholders, and other constituents. You will learn more about developing a shared vision for change in Chapter 3.

READ MORE ABOUT IT

If you wish to learn more about the rhetoric of association and how it can be used for legitimacy, take a look at Suddaby and Greenwood's work.

Suddaby, R., and R. Greenwood. 2005. Rhetorical strategies of legitimacy. *Administrative Science Quarterly* 50: 35–67.

CHANGE READINESS ASSESSMENT

Summary: When you consider launching a change project, you'll need to assess your own readiness—as well as the readiness of your team, your committee, your department, your campus—to do the work.

People get ready, there's a train comin'

—Curtis Mayfield

So now you know a bit more about how you can prepare yourself to reach your change goals. You can use one, some, or all of the tools to move yourself, and your project, forward. The foundation of the book is designed to help you equip yourself for the work that lies ahead. Before you jump in, however, you might ask yourself if you are ready to start. It's a lot like taking a baseline assessment if you want to change some aspect of your current health, or a baseline strength test if you want to work on increasing your ability to bench press free weights. Like these assessments, there is a way to assess your change readiness as well.

By sharing some of the items from the Change Readiness assessment, developed by David Holt and his colleagues (2007), you can gain insight into where you are with your change project and your current level of change skills, what tools you already possess, where you may need to call in additional help, and so on. The assessment is also useful if you wish to determine the readiness of your team, your committee, or your department. After you complete the assessment, take some time to review your results and reflect on what they may mean. Even discussing the items with your team and determining how they think and feel about change can be a beneficial first step. The tool's items are grouped into five areas of change readiness:

- Feels confident in one's ability to make change: items 13, 14, 19, and 22
- Sees a need for change: items 2, 4, 23, and 25
- Sees personal benefits accruing from the change: 1, 8, 9, 11, 12, 15, 17, 21
- Sees benefits to the organization resulting from the change: 3, 5, 6, 10, 16
- Believes senior leadership supports the change: 7, 11, 18, 24

Identify the top challenges or gaps that emerged, so that you can plan to address them, or at least not be surprised when they emerge along the way. In the same way that you pack emergency tools when you take a road trip—like a set of jumper cables, a can of Fix-a-Flat—you can be ready for change when you bring along what you need, rather than relying on the kindness of strangers who may or may not stop to help you by the side of the road.

Used with permission from the publisher.
Score all items on a Likert scale 1= strongly disagree to 5 = strongly agree.

---

1. In the long run, I feel it will be worthwhile for me if the organization adopts this change. 1– 2 – 3 – 4 – 5

2. It doesn't make much sense for us to initiate this change. 1– 2 – 3 – 4 – 5

3. I think the organization will benefit from this change.   1 – 2 – 3 – 4 – 5

4. Management has sent a clear signal this organization is going to change.   1 – 2 – 3 – 4 – 5

5. This change makes my job easier.   1 – 2 – 3 – 4 – 5

6. When this change is implemented, I don't believe there is anything for me to gain.   1 – 2 – 3 – 4 – 5

7. My past experiences make me confident I will be able to perform successfully after this change is made.   1 – 2 – 3 – 4 – 5

8. My future in this job will be limited because of this change.   1 – 2 – 3 – 4 – 5

9. This change will improve our organization's overall efficiency.   1 – 2 – 3 – 4 – 5

10. I am worried I will lose some of my status in the organization when this change is implemented.   1 – 2 – 3 – 4 – 5

11. There are some tasks that will be required when we change that I don't think I can do well.   1 – 2 – 3 – 4 – 5

12. I do not anticipate any problems adjusting to the work I will have when this change is adopted.   1 – 2 – 3 – 4 – 5

13. When I set my mind to it, I can learn everything that will be required when this change is adopted.   1 – 2 – 3 – 4 – 5

14. I have the skills that are needed to make this change work.   1 – 2 – 3 – 4 – 5

15. This change matches the priorities of our organization.   1 – 2 – 3 – 4 – 5

16. This organization's most senior leader is committed to this change.   1 – 2 – 3 – 4 – 5

17. The time we are spending on this change should be spent on something else.   1 – 2 – 3 – 4 – 5

18. When we implement this change, I feel I can handle it with ease.   1 – 2 – 3 – 4 – 5

19. Our organization's top decision makers have put all their support behind this change effort.   1 – 2 – 3 – 4 – 5

20. I think we are spending a lot of time on this change when the senior managers don't even want it implemented.   1 – 2 – 3 – 4 – 5

21. There are legitimate reasons for us to make this change.   1 – 2 – 3 – 4 – 5

22. Every senior manager has stressed the importance of this change.   1 – 2 – 3 – 4 – 5

23. There are a number of rational reasons for this change to be made.   1 – 2 – 3 – 4 – 5

24. This change will disrupt many of the personal relationships I have developed.   1 – 2 – 3 – 4 – 5

25. Our senior leaders have encouraged all of us to embrace this change.   1 – 2 – 3 – 4 – 5

Once you have completed the tool, you may be asking, what should my score be? I suggest that you use your responses (and the responses of your team, if you are using the tool as a group) to reflect on what your answers suggest about your readiness for change, rather than seeking out the "right" score:

- Are there specific gaps or areas that you will need to address before you start?
- Is your team ready? Should the membership of the team change to provide adequate support for the project?
- Was there anything surprising about your results? Are you more prepared to do change work than you thought?

Discussing individual answers within the team can also provide some important insights. If you use the assessment with your team, your committee, or your department, be sure to allow for individuals to share their perceptions and gather their feedback. Taking time at the start may often give you a preview of challenges that may arise in the future!

READ MORE ABOUT IT

You may be interested in learning more about Holt's work in the area of organizational change that he developed out of the assessment tool.

Holt, D., A.A. Armenakis, H.S. Field, and S.G. Harris. 2007. Readiness for organizational change: The systematic development of a scale. *Journal of Applied Behavioral Science*, 43(2): 232–255.

And you can check out the numerous recordings of the Curtis Mayfield classic. The performance by Reverend Al Green, joined by Wanda Neal and Linda Jones, is inspiring!

INSTITUTIONAL CONTEXT AND "THE BATTLE OF SAN ROMANO"

Summary: In this section, you will perform a scan of your campus environment. The purpose of the scan is to identify issues that may be leveraged in support of your project or could pose threats to your project before it begins.

In "The Battle of San Romano," the Italian painter Paolo Uccello depicts opposing Florentine and Sienese forces facing each other in a fight for control. In the foreground, the battle rages. Niccolò da Tolentino, leader of the Florentines, brandishing his sword, sits on his white horse. Banners denoting opposing sides furl and flap in the wind. Men fall face down in the dirt, while above them, horses strain against their riders. The energy of the battle scene is conveyed through the postures of the horses, the attitude of the soldiers as they aim their lances, almost as if we can hear the thunder of the animals and men as they fall and die. While the foreground of the painting is arresting, what also captures my attention is what is happening behind the horses and their riders. The viewer's gaze moves up to the center of the painting only to find a bucolic pastoral scene. There, men and hounds pursue hares across open fields. A farmer tends to wheat that is ripening in the field. Warfare and rural pastimes, two scenes shown side by side. And participants in one location are oblivious to those in the other.

- Does the depiction of the battle remind you of something going on with your campus currently?
- Who is the Niccolo da Tolentino on your campus?
- Who are the individuals on horseback and who is chasing rabbits?
- What better picture exists of a conventional college or university campus in the early 21st century?
- How appropriate is this image to the circumstances in which each of us works to move our change projects forward?

The circumstances depicted in the painting are easily adapted to contemporary academic contexts. Consider the change project of Professor Nicola Tolentino, a distant relation of Niccolo, who is immersed in her change project: revising the customary approach to teaching organic chemistry to undergraduate students in her department. Like her forbearer in Uccello's painting, each day she must arm herself for another skirmish. On Tuesday, she presented her proposal to the college curriculum committee. By Friday, she received emails from colleagues who encouraged her to continue forward. Each day, she feels consumed by her project, knowing that she is working toward the right goal, to improve her students' learning, but also keenly aware that she must fight battle after battle to make the change she believes in.

But also like her predecessor who fought in the Battle of San Romano, Dr. Tolentino is unaware that, just beyond the edges of her sight, the rest of the campus is tending to its own interests and concerns, completely oblivious to the hand-to-hand combat she must win if she has any hope of implementing her new pedagogy. From time to time, she looks up from her desk and scans the campus quadrangle where students toss the Frisbee or lounge on the grass, faculty colleagues passing by each other as they hurry on their way to class.

- Why, she asks herself, is her project not getting noticed?
- Why are those colleagues on the other side of the quadrangle content to tend to their own fields, to chase another rabbit?

In Uccello's painting, the soldiers in the foreground never take a moment to consider the scene in the background. What if they, like their neighbors, gave up arms and beat their swords into ploughshares? What if, instead of killing each other, they joined the hunt to kill hares? Should Dr. Tolentino raise her head from her own concerns in order to consider what lies just beyond her own field?

The work of an academic change maker requires devotion and energy, that is true, but it also requires attention, and not just to the change maker's own project. Lifting your head from own project to scan the environment around you has positive benefits. Not only do you increase your awareness of the concerns and challenges that are current on campus, you can identify potential partners whose challenges are not your own but who may be at work on issues that intersect. For example, Dr. Tolentino may not find a colleague who also wishes to teach organic chemistry in a new way, but she may find three other colleagues who are challenging the status quo on other required courses. And changing required courses can often make allies of the change makers who must face off against change opponents in contested meetings of the curriculum committee.

Put the Tool to Work

In order to understand the issues and challenges that preoccupy members of your campus community, a good first step is to perform an organization environment scan. Using a few easily accessed information sources (i.e., the campus intranet, a committee meeting, the student newspaper, a department seminar, an external speaker invited to campus, etc.), compile ten issues that appear to be garnering attention among students, staff, faculty, and administrators. For example, a quick environment scan at Dr. Tolentino's campus would result in the following list:

1. Hiring faculty for diversity and inclusion
2. Rising costs of healthcare premiums for staff
3. Availability of financial aid targeted to first generation students
4. Academic performance of students in first-year courses, particularly in math
5. Loss of commuter parking lots due to construction of the new nursing school building

Granted some of these items are not central to the pedagogical change that Dr. Tolentino wishes to enact, but there is potential in the concerns regarding items 3 and 4. The next question is, who is most involved with the financial aid question and who is collecting data about academic performance? Both of these issues have the potential to connect with Dr. Tolentino's work, and her work on an institutional scan can help identify relevant issues and the individuals who are working to resolve them.

So what will emerge when you perform your institutional environment scan?

> **Your environment scan: what issues is your campus currently contending with?**
> 1.
> 2.
> 3.
> 4.
> 5.
> 6.
> 7.
> 8.
> 9.
> 10.

READ MORE ABOUT IT

As I promised in the introduction to this book, each tool has its origin in research literature, but the research field relied on may not be one that you regularly consult. This is especially true for the environmental scan tool. If this related field is of interest to you and you would like to read more, check out these two pieces from the information science and information management fields (Albright 2004; Choo 1999). Happy reading!

Albright, K.S. 2004. Environmental scanning: Radar for success. *Information Management Journal* 38(3): 38–45.
Choo, C.W. 1999. The art of scanning the environment. *Bulletin of the American Society for Information Science and Technology* 25(3): 21–24.

YOUR ROADMAP FOR CHANGE

Summary: The road to change is a challenging one, so you'll plan your route before you start.

> I've always been fascinated by maps and cartography. A map tells you where you've been, where you are, and where you're going — in a sense it's three tenses in one.
> —Peter Greenaway

In 1539, the Swedish clergyman Olaus Magnus produced the *Carta marina et descriptio septentrionalium terrarium*, which translates from the Latin as *Marine map and description of the Northern lands*. Magnus' creation was the first map of the Nordic countries to provide specific place names and details, some based on scientific treatises current in his time, some based on his own observations and observations provided by sailors who traveled these waters. The details are what make this map a work of art and a testament to the ways in which fact and science fiction could be combined without readers raising objections.

Sitting in the middle of the map is an area that Magnus wants to document and warn us from entering. In this spot in an enormous ocean, ships have encountered difficulties, although the causes may not be scientifically verifiable, like the red sea serpent that Magnus depicts just below the waves. The message, however, is clear: don't venture into this region unless you are willing to risk almost certain destruction.

Maps have always been important for human cultures. Whether you needed a map to remember where you buried a tasty piece of dried mammoth meat or a map like this one to help avoid the sea monsters inhabiting the depths of the Atlantic, a map is, as Peter Greenaway says, three tenses in one. Change makers also need good maps, whether, with their change project, they are exploring unchartered territory or following a more established track. The process of making your own roadmap will help you identify several important features of your work: the milestones you believe you will reach along your way; the estimated time you will need to reach each milestone; and the impact you hope to make on your stakeholders.

But as Olaus Magnus and every other map maker in history have known, making a map requires planning. In order to figure out where you are going, you need a way to determine your route, with options to explore a less traveled route, or find a way to get around an obstacle. And you need a visual of the journey in total, a means to understanding the trip in a single glance.

If you were to make a map of the change journey you wish to embark on, you might draw an X to make your starting point, and perhaps a Z to denote the place where you'd like to end up.

- But what lies in between?
- Which routes are available to you?
- How will you know if you are still on the right road?
- At what point will you realize that you have lost your way?
- How can you get back to the right path?

That's why we will create the roadmap for your change journey, to help to see how you will get from start to finish.

This reflection on maps provides an introduction to a two-part tool for planning. As a starting point, you will draw a map freehand that visually depicts your journey toward the future state that will emerge as a result of your project. In the second part, you will turn that visual depiction into a concrete plan that addresses important stages of your project. This second "map" will serve as your Roadmap for Change for your project, something that you can return to again and again to ensure that you are keeping to your route, no matter how many obstacles, diversions, sea monsters, or byways you encounter along the way.

PUT THE TOOL TO WORK

Part 1—Your Roadmap

In making Your Roadmap for Change, you can engage both the fanciful aspect of mapmaking in the style of Olaus Magnus and the concrete aspect of mapping to devise a roadmap, or plan, for your change project. In order to create the visual map for your project, collect a large sheet of blank paper and colorful drawing markers, and clear out enough space on your desk to draw. Denote the starting point of your project on the paper, as well as the end point, then draw the route that you anticipate you will follow to get from the start to the finish. Of course, you are completely in your right to draw a straight line from point X to point Z, but the map can also be the place to identify the challenges and successes you anticipate you will encounter along the way.

- Consider adding a perilous river crossing as a metaphor for an obstacle you know you will need to cross.
- You can also include possible successes along the way, as if you can already see the mountain peak you will summit when you receive recognition for your project.
- If you like, you can add a few trolls behind a boulder to remind yourself that there will be individuals who may try to thwart your progress.

No matter what you include as summits and challenges along your way, try to include several milestone markers that represent specific stages of your project, with a notation of when you think you will get to that point. For example, your first milestone could be a meeting of your college's curriculum committee where your proposed project may be reviewed and voted on. Take a moment to add to that milestone the timeframe in which consideration by the committee will occur. You may find it useful to identify several milestones and their associated timeframes in order to help you recognize when you are one-quarter, one-half, and three-quarters of the way to your final goal. You should also consider adding the tools you have already acquired to this fanciful map. You might, for example, adopt the metaphor of the Indianapolis 500, with Emerging Opportunities, Change Readiness, and Environmental Scan tools depicted as a series of lights or flags that indicate that you are ready to start the race. Change makers, start your engines!

Part 2—Your Plan

While your Roadmap serves as a visual depiction of your project journey (and hopefully you are taking advantage of your artistic skills to make it colorful and inspiring), you also need a version of your map that is more traditional. Granted, you could spend quite a bit of your time devising a project plan in any version of project management software. At this stage, however, focus on the significant milestones that will help you reach your final goal. It may help to consider each milestone as it impacts four different domains.

People: who do I need to talk to, collaborate with, enter into a partnership with, in order to reach this milestone?

Resources: what resources do I need, how much, by when?

Stakeholders: how will specific stakeholders (students, faculty, staff, community members, etc.) experience each milestone and how will their day-to-day existence change because we reached this milestone?

My experience: what will look different to me when I reach each milestone, what challenges or dangers can I anticipate with each stage of my project, how will I be different?

Now that you have two planning tools—your visual map and your written plan—take a moment to review your work. Post the visual map near your workspace so you can look at it every day. The map is meant to inspire you to move forward on your project, even as it serves as a reminder that, like Olaus Magnus, you may need to consider alternative routes if you are ever waylaid by a half-submerged creature of the deep.

READ MORE ABOUT IT

Although Olaus Magnus wasn't a sailor himself, he talked with mariners who explored the wild and untamed regions of the globe. And you can see his map when you visit the University of Minnesota James Ford Bell Library.

CHANGE MAKER INTERVIEW: DR. JEREMI LONDON

At the end of each chapter in this book, you'll be introduced to a STEM change maker through an interview. The purpose of these interviews is to help you see the change tools in the context of real-world change work, highlighting the fact that many change makers employ these tools in different contexts and for different projects. In this interview, Dr. Jeremi London, associate professor of engineering education at Virginia Tech, gives us insight into her motivations to be a change maker and how she approaches the change efforts she works on.

During the interview, I asked Jeremi to offer some advice to change makers. I thought her comment was particularly helpful:

> Give yourself permission to dream out loud. Sometimes we have dreams, and we just keep them to ourselves or we doubt ourselves or we come up with a thousand reasons why that can't work. But I have come to realize that there's only one you. My daughter is one, so I'm familiar with Dr. Seuss books, and the phrase I repeat a lot is, "There's nobody you-er than you!" There are certain things that only you are fascinated by, and there are certain things that only you notice, and there's certain ways that only you would do things. Nobody's you-er than you. So there are certain dreams that only you will have, and it's okay to dream out loud. So I'd say give yourself permission to dream out loud.

That is advice every change maker can benefit from!

Dr. Jeremi London is Associate Professor of Engineering Education and the Instructional Innovation Lead at the Virginia Tech Innovation Campus. London's commitment to bridging the gap between research and practice has led to meaningful student outcomes and national leadership roles, the most notable of which was Chair of the American Society for Engineering Education's Commission on Diversity Equity and Inclusion (CDEI) during the professional society's *Year of Impact on Racial Equity* (2021–2022). Her scholarly interests have been supported by over $6 million in funding and resulted in over 100 peer-reviewed articles, best paper awards, and keynote addresses. London's most notable award, an NSF CAREER award titled "Disrupting the Status Quo Regarding Who Gets to be an Engineer," blends her interests in the study of the impact of academic change with the study of broadening participation of underrepresented groups in engineering. London holds BS and MS degrees in Industrial Engineering, and PhD in Engineering Education, all from Purdue University.

JL: There are probably two or three big things that come to mind when I think about what am I doing in the world of change or thinking about change. Two of them are on the research side and one is on more of a service side. As part of a faculty role, I think of my life in terms of teaching, research and service. As teaching-related, this past semester I designed and taught a new course on the research life cycle, and as part of the research lifecycle, I was talking about how you link research and practice and how you make change in the classroom. This was the first time I asked graduate students to think about how their research interests could make a difference in some practical way. I was trying to use my research insights and translate that into a course.

On the research side, there are two big change-related activities I can think of. First, I'm fascinated by the dysfunctional relationship between research and practice. In engineering education, we do a lot of research, and we're awfully busy in the world of practice. But those worlds don't always collide, and so we aspire to see change happen in terms of teaching students or advising administration. There are all kinds of ways in which we desire change, and I think that there's a lot of scholarship that could inform change in those contexts, but those worlds don't always collide, so I think that there's a really dysfunctional relationship between them.

And so one of my primary research interests is, what makes that relationship dysfunctional and how do you bridge the chasm between them? I was recently awarded a NSF CAREER grant, which I think of it as the Olympics. When I'm describing it to people who are not in the academy, I describe these grants as winning in the Olympics, because you're the athlete within a event, you usually do it when you're relatively young, and you're only eligible if you're early in your career as a faculty member. And you really have to be among the best in your field, your ideas have to really rise to the top.

For my grant, I have been specifically awarded to study the change strategies of a few exemplary institutions and specifically colleges of engineering that have done a phenomenal job of recruiting, retaining, and graduating black and brown students over the last decade. I want to understand their change strategies. I'm using a framework by John Kotter who has outlined a theory of change in three big categories: envisioning, implementing, and institutionalizing. Based on those three stages, I do not think that the schools that have become these exemplars achieved that status by chance, I just refuse to believe that.

So I want to know, long before they were among the best, what did they envision? What did they aspire to be in terms of their recruitment, retention, and graduation of black and brown engineering students? And then what did they implement? What actual changes did they put in place? And then how have they made it institutionalized such that even with people turning over and new cohorts of students, there's all kinds of stuff about their systems that are not the same, that's not steady, but yet they continue to be "the best" as designated by the American Society of Engineering Education by the numbers. ASEE publishes statistics each year about who are the top producers of black and brown engineering students. So I want to know, what have they institutionalized such that they consistently are named among the top? These top five are University of Maryland, Baltimore County (UMBC), Morgan State University, the University of Maryland College Park, University of South Florida, and Florida International University.

And I need to include my work in service, because I am the incoming chair of the ASEE Commission on Diversity, Equity and Inclusion, and I will be leading the Year of Impact on Racial Equity. About five years ago or so now, ASEE had a year of action on diversity and in some ways this upcoming year is a follow on to that. We're trying to go beyond action to impact, and we want to see some observable change happen. There are three pillars in the Year of Impact initiative; one is focused on student organizations, so realizing that engineering culture is manifest in students' interactions with one another. Oftentimes we talk about administrators and faculty, but the one constituent group that I feel like we haven't engaged much are students and one of the places where students congregate is within student organizations. We're challenging student organizations to think about the engineering design process, but to focus on a diversity issue within their student organization, and how could they make change using engineering design. The second pillar relates to administrators of engineering faculty and administrators. All of them have their own quirks and challenges, but for that one, we're asking them to revisit their admissions and their promotion and tenure policies, not that they will change them in a year, but to identify or spot equity issues. The details are still being worked out, but we're working with the Engineering Dean's Council to specifically look at admissions and equity, because those are two policy points that determine who gets to be in engineering as a faculty or as a student.

The third and final pillar relates to K-12. We're partnering with the P-12 Commission, which is also a part of ASEE, to try to get more black and brown children engaged in engineering outreach efforts. So realizing that there are all kinds of ways to expose people to engineering, but there are certain groups who tend to participate and some that tend not to. As part of increasing awareness and interest in engineering among the most marginalized groups, that pillar will address increasing representation of black and brown children in the existing outreach efforts.

JW: What inspired you to take on one or more of these projects?

JL: A bit of craziness, a bit of insanity in there. There are lots of ways I could answer this question, but I'm going to blame it on being in industrial engineering. I am an industrial engineer by training, and whatever the series of events in my life that caused me to choose IE, there's something about the way that I see the world, the way that I wish the world worked. Industrial engineers think about systems and about how well a system's working or how well it isn't working, and they usually think they can do something about it. Typically, industrial engineers are focused on manufacturing processes or developing widgets, and these were the example I saw in almost every one of my IE classes.

Now I'm not interested in widgets, but I'm going to focus on seeing that systems work well. That is something that undergirds everything that I decided to say "yes" to. I always ask, how big of a problem is it? Now I wonder, Jeremi, did you pick all of the big problems to do all of them at one time? When I look at a problem, I ask myself, why is it this way? Is it more of a systemic issue and something not working so that we could do something about it? Even if we make one tweak, for example, we will see the impact of that reverberate throughout the system. So I'm going to blame my IE mindset on why I opt to do certain stuff. Now why do I sign up for certain stuff? Why do I propose certain stuff? Because I have this view of the engineering education as a system and realizing that if you pull on the right levers, sometimes it takes a while to figure out what the lever should be, but pulling on certain levers can lead to changes at the local level and beyond.

JW: That's a great answer and it's something that I'm hearing from other people working on change projects. They tell me, "I see something, it's a problem, I want to go in and I want to work on it." I think that's very common. But even as you do that, you are a relatively young person early in their career, as the NSF award indicates. Do you see any potential risks or have you experienced any challenges from people who say "Hey, Jeremi, it can't be you who works on this," or "It isn't the right time."

JL: When I did my dissertation seven years ago or so, I had this guy on my committee named Karl Smith, and I referred to him as the old wise guy on my committee. Karl said, "You chose a wicked problem." It's a fascinating problem, like when I was specifically focused on the impact and the relationship between research and practice. These wicked problems have certain characteristics: they're really messy, there are a lot of moving parts, the constraints change, there's no one right solution.

Picking a wicked problem inherently is difficult, and one of the things that make it hard is realizing there is no roadmap for you to follow. Sometimes there aren't even people that you could talk to about it with, because you feel like you're in a world on your own. You're not sure who else is fascinated by this problem. Who else has tried to delve into it? There is a lack of community, a lack of a charted path, and sometimes it's just hard. You feel like you're making progress, or you hope you're making progress and then you ask yourself, "Am I making progress?" You wonder, are you headed in the right direction? Is this going to move the needle? There aren't clear metrics, I guess, is the way that I probably described it.

In other contexts, there may be easier landmarks or signposts to say you're headed in the right direction, but when you choose a wicked problem, you don't always have that, and I guess you have to rely on other indicators that things are working, that things are successful. The nature of a risk is that you don't know the outcome. You ask yourself, "I don't know if it's going to work. Why did you propose to work for a whole year on impact during a pandemic?" I mean, I'm only a chair of CDEI for one year, I can't help that it overlapped with the year of the pandemic. Now, how are you going to run a whole society level initiative from your home office? How are you going to mobilize a whole community without face-to-face meetings or conferences?

There's risk, is what I'm trying to say, there's all kinds of risk. One of my favorite memes (I'm a sucker for a good meme) "What if I fall? But, oh my darling, but what if you fly?" And I feel like there are these critical moments in life where you're at a juncture where you ask "Well, what if I fall?" Then there's this other follow-up question right after "Well, what if you fly?

JW: We need t-shirts for that one.

JL: For real, I love that meme, I came across it when I was doing my dissertation, when I first picked this wicked problem, and I asked myself, "But what if you fall?" But it's the same now embarking on the CAREER grant or this year of impact. I have always just reminded myself that the quote doesn't end there, it's what if you fly? And realizing that it's okay to fall sometimes, it's not the end of the world.

JW: Let me back up just a little bit, because I think your sense of not having a clear path, not having metrics, not having landmarks, I think that's common to a lot of people who are taking on a change project in different contexts. You feel like you're doing this on your own and who cares? So how did you find your people, your community, the ones who say, "Yes Jeremi, you're on the right track," or, "Oh no, you've got this." How did you find these people?

JL: I have started to realize that I need layers of community, so it took a while to find the people who are specifically focused on the content area that I'm really focused on and to even start to use the same jargon so I could find them. I had to identify the right keywords in order to figure out the conferences where these people gather. And I started to realize that the majority of the people were outside of the US, the community that studies impact as a construct. So once I started to find my people in terms of impact and realized that I needed to travel to go to the impact conferences that are usually abroad, that was helpful to know. I wasn't missing them, they're just not here in the US, and I have to go other places.

So starting to identify the keywords and the right conferences, that was how I started to find my people in the impact space. But I don't care about impact in the abstract; I care about impact in the context of engineering education. So I knew I needed an impact community or at least engineering education community. In that case, like finding those people, I started to realize that there's an expression in the impact world where impact is described as a terrain in which there are multiple groups of people who are engaged in impact work. And once I realized that there were multiple groups of people on the terrain who talk about impact slightly differently while they're talking about a similar phenomenon, then I said, "Oh, when I'm in this context, this is how I have to talk about it in order to find my people." And once I started to find my people, then those folks would say, "I think you're onto something, Keep going." Or they would say, "Hey, come give a talk." And I'd ask, "You want to hear what I have to say?" So those kinds of things will keep you going and help you to find people that care about what you're interested in and about what you have learned.

Those are the things that keep me going, realizing that I need multiple layers of community. There are some that are specifically focused on impact and they're focused on their own context, but we can talk about impact in the abstract. In the world of engineering education, we talk about translating research to practice or evidence-based strategies as a way of understanding impact. Once I realized what impact looks like in this world or that one, then it made it easier to find where my people are in those places.

JW: I think your realization points to the importance of how you're using communication and adapting your language for different groups and contexts. I think that's really a helpful piece of advice as well.

JL: One thing that I have come to realize over time is that I could not just study impact. That's where my interest in change started to come about. Once I realized that impact is typically the final result that we're fascinated with, I saw that there's a lot of stuff that happens and leads up to the final result. And so that's where I've started to become interested in how change happens, like what are the things that enable and impede change.

JW: And as you're looking across that change landscape or terrain, are you seeing some trends in terms of resistances, the things that get in the way?

JL: I can't say that I am an impact on or expert on change yet, but I had to know enough to be dangerous in order to write a National Science Foundation CAREER proposal. So what I've started to realize is that there are some paradigms around how change happens and with each paradigm, there are different forms of resistance. Like there's times when some may think that change happens organically, that it just comes about by some whimsical set of events, and then somehow we magically end up in a totally different place. There's some change that is planned, which is the camp that I'm in when I think about the CAREER project. So with each of them, there's other paradigms changes, but with each of them, let's say we go with the more whimsical, organic approach, in that case, some of the barriers or resistances are that you don't have as much control about where things are headed or you don't even know if you headed in the right direction. And in that case, it may be harder to pick who is leading this change, and should we follow their direction.

But then on the other hand, if it's planned, you got to try to motivate people to buy into a vision and then you have to come up with some clear metrics to know whether we're headed in the right direction, like mobilizing existing resources to go towards some new effort. I feel like on either side, there are challenges and resistance, ways in which change can be difficult to attain. I've come to realize that although change is extremely common, and I want to say something more sophisticated than "hard," but it is difficult and constant. People say the only thing constant is change and yet it is one of the biggest hurdles to getting things done. It's really an interesting phenomenon, it's one of the paradoxes, that change is so normal.

JW: I mean, we went from face-to-face classes to everything online almost overnight because of the pandemic.

JL: Yet for years we talked about maybe opening up new pathways or creating an online program or cyber learning, which was my first research passion. Back then, there was every possible reason why that wouldn't work or why we couldn't possibly do that. Then when you're forced to do it, it's like magic.

What happens if you're forced to? I may start asking people that question: when you say you can't do it, well, what happens that you are forced to do it?

JW: What did you learn from that and what can you take going forward? How can you make things better rather than just saying, "Oh, I want things to go back to normal." I think the "normal" is what the problem was.

JL: That's right. But something as disruptive as a pandemic could force you to go back to the drawing board. If there's anything good that comes out of this, then it's giving people permission to dream out loud and rethink what they do and how they do it.

JW: Right. You talked about your background as an industrial engineer and how it prepared you for looking at things in a particular way. Do you think there's any other part of your background or something about you in particular that makes you that person who's always going to be looking for that challenging project?

JL: So I use IE as a cover, but I attribute a lot of my thinking to the mentoring relationships that have shaped my life. There is a handful of people, I call them my "Fave Five." These mentors collectively have given me big shoes to fill, and I'm pretty sure that informed some of my ambitions. They are my middle school teacher who is now a superintendent, my high school principal who now runs a city-wide program in Chicago, and a couple of NSF program officers who have a global view. I can't help but think of the Fave Five that I look up to. I also feel like I have a pretty supportive family who say, "Whatever you decide to do, go for it." But I think that it is my mentors who have given me a vision for how big you can dream, like realizing that there really is no limit. Now, whether you should go that far, is a different question.

JW: It's up to you.

JL: Yeah, but imagining what's possible, like between the five of them, I have seen even them do some incredible things. Like the NSF mentor that I speak about, helped start the I-Corps for Learning program. There was already an I-Corps program where you could imagine how a research idea could have some entrepreneurial dimension. But realizing that the same entrepreneurial dimension could happen in the learning space, that's pretty radical. So I can't help but imagine the role or the influence of those, not just role models, but mentors and realizing that there are really big, fascinating problems that need to be addressed and why shouldn't you be the one who takes them on.

JW: This year of impact project is going to bring lots of different people together. Do you have a strategy for collecting these people and helping them work together effectively?

JL: The Year of Impact officially kicks off at the ASEE conference, and it's organized in a tiered strategy. CDEI has a leadership structure in which there are always three active chairs. There is the current chair, there's a past chair, and there's an incoming chair. Currently I'm the incoming chair and then starting at the conference, I will become the active chair and then next year I will be the past chair. The organizational strategy is three pillars, with each chair overseeing one pillar. We all meet weekly, so we'll share updates that way.

CDEI has quarterly meetings so any member of ASEE can join the CDEI quarterly meetings. These are generally people who are interested in diversity, equity and inclusion and want to know what's happening at a society level, not just within their division, in terms of diversity. I've also created an interest survey. So I've leveraged the CDEI quarterly calls to share the interest survey in which people could get excited about a pillar. Fifty-five people have signed up to be on the task force for the year of impact. As part of the interest survey, I asked them to specify which pillar they were most excited about, and I collected other data through the survey. I asked them to tell me about some demographic characteristics because I would like the makeup of the task force to be diverse, then I also asked them, which of the three pillars are most excited about. It's a way to tap into their motivations and not assigning people a pillar, but asking them to identify their preference. That way, there's autonomy built in. Also, I asked them when might they be available, and I gave them the option to say, "Do I want to be involved in the planning?" Some people say, "I can give you some time in the summer, but please don't talk to me in the school year." Or, "I don't give a flip about the planning, but I'll help you implement." And then there's the tail end and celebrating at the next conference. So there are some people who said, "If you do the work and stuff, I can help when you're at ASEE 2022, whatever happens there, I can help with that."

I gave people options in a lot of ways. It's like, "Which pillar were you interested in? And then when would you like to be involved?" And so between those two main questions, the task force of 55 is now distributed among the three pillars. Each of the chairs will oversee their group, so I'm in charge of the one related to the student organizations. Now we're starting to kick off our task force activities. That is the initial way in which I am trying to rally people around the year of impact. Then at the 2021 ASEE conference, the incoming president of ASEE will announce as part of her presidential address at the end of the conference the year of impact. Having her leverage her platform to talk about the year of impact, that will be our last big push for the interest survey in getting people on the task forces. So that is how I am rallying people from my address to be involved in a year of impact on racial equity.

JW: Fantastic. So was there something that you thought I would ask you about during this interview, a question I would ask that I didn't, that you'd like to respond to?

JL: Honestly, I don't know if I knew what I was signing up for. I don't know if I had expectations. I was like, "You're my friend and if my friend is writing a book then whatever you want to talk about is what I'll talk about." That was really my reason for saying yes. A lot of my friends are interested in cool stuff, so you just like whatever their thing is, that was the real motivation for saying yes, and I'm also fascinated by change in general.

JW: The way you described the different projects you're working on, you seem to be saying that you may not have the exact strategy worked out, but you know what direction you are going. You're going to roll with it, you are flexible…

JL: Yes, I do think that a part of my personality, I know I have labeled my mindset as IE, but it's really how I see the world, as well as the influence of my mentors. I am fascinated by serendipity. I look forward to things not panning out exactly the way that I thought and being surprised by it. I am fascinated by the journey, I guess. When I decide to embark on certain endeavors, I am genuinely curious about how this is going to turn out, how it's going to end up. That uncertainty motivates me to keep going in that initiative and to try new stuff, because I'm just genuinely fascinated by what could happen.

JW: And your department at Virginia Tech is supportive of the work that you do in this project and the others, is that correct?

JL: Yes, that is true. That also helps with being able to do cool stuff. It's having a supportive leadership team and colleagues, that helps with doing interesting things. It's hard to imagine being in a place that is not supportive. So I was talking to a mentor recently about what should I be dreaming about next in terms of my career moves, and I was thinking, "I think I could work anywhere." And they said, "No, you can't work anywhere, some places it's harder to get stuff done." And I guess I've been blessed or fortunate that that hasn't been my reality. The fact that I haven't experienced a lot of resistance also influences me to think I can reach for the moon, and it's how I land among the stars.

JW: I think you need to have that very strong, positive, and confident attitude, to tell yourself, "If this doesn't work out, I'll just check the work and try something else." So that makes me wonder if you had advice to give to the people who are going to use this book. They're looking for some support, I think they're searching for a community.

JL: I guess one piece of advice is to give yourself permission to dream out loud. Sometimes we have dreams, and we just keep them to ourselves or we doubt ourselves or we come up with a thousand reasons why that can't work. But I have come to realize that there's only one you. My daughter is one, so I'm familiar with Dr. Seuss books, and the phrase I repeat a lot is, "There's nobody you-er than you!"

There are certain things that only you are fascinated by, and there are certain things that only you notice, and there's certain ways that only you would do things. Nobody's you-er than you. So there are certain dreams that only you will have, and it's okay to dream out loud. So I'd say give yourself permission to dream out loud.

JW: I also liked the way you characterize it, that you Jeremi, me Julia, we would not approach a problem the same way, we would not have the same talents to

bring to it and it would be common that we say, "Oh well, I'm sure somebody else…" No, nobody else would do it this way. No one else would do
the year of impact the way you're going to do it.

JL: That's right. In the words of one of my favorite teachers or professors, I can't
remember if it was Dimitri Evangelou or Ruth Streveler, they both co-
taught this class. But between the two of them, they affirm one another.
They say "all research is autobiographical," and I love that, I have always
remembered that. And the reason why you specifically want to study that
thing, says something about you.

There's a reason why I study certain stuff. You can learn a lot about me
from what I study and the same, like when I look at the things that fasci-
nate you, it tells me something about you. So be mindful about the things
that you give your time and your life too, because what you do is autobiographical, at least the things you study or the change projects that you
decide to take on say something about you and there's nobody you-er than
you,. Give yourself permission to fly. I don't want to dismiss how hard it
is to do things we're doing, it's okay to acknowledge, I guess, the challenge of doing hard stuff, but don't stop it, acknowledging the challenge,
figure out what would it take to be successful. So it was last week that I
said out loud about the Year of Impact. "This is big." This is really big.
Now mind you, this idea was proposed to the ASEE board back in
November, it got approved in January, it's has been months, we've been in
it. And now that people have responded to the interest survey and it's
marshaling 50 people, I was like, "This is big." But it's okay to feel that,
it's okay to acknowledge when something may be, like a whole elephant,
you eat an elephant one bite at a time, but it's okay to feel that, but then
you start to realize like, "Well then what do I need to be successful?"
What would it take to be successful?" I don't want to just diminish the real
feelings, the real fears, the real acknowledgement that I might be a butterfly, but I don't think I'm falling, I think I'm a butterfly.

JW: Just flap your arms harder and faster!

JL: So my piece of advice is, give yourself permission to dream out loud because
nobody's you-er than you, and it's okay to sit with the heaviness or the
weight of what you decide to embark on, but don't stay there. Figure out
what you need in order to be successful.

JW: Very wise, very good advice. Are any questions you want to ask me?

JL: Can you tell me why you care about change?

JW: I think even though I am not an industrial engineer, I've always been pursuing academic change. I arrived at Rose-Hulman in 1992, and early on,
I started a change project. Rose-Hulman adopted a one-to-one laptop program back in the day when laptops weren't all that great. I thought, "Gee,
why don't we have a way for the humanities and social sciences to use the
laptop technology? What's software is out there?" And I worked to get
adoption of a composition program for writing.

Looking back, I see that I made some mistakes with that project, but
boy did that experience influence how I did the next change project, which

was also a technology integration, using tablet computers with a software. And I didn't stay in my department. That's when I started reaching out beyond my department into other areas to find like-minded people, the people who were interested. And that just started a whole career of, "Oh, I'm going to do this thing," and "I think we need to have a change workshop on our campus." Yes, why not do that?

JL: That's what I mean by dream out loud. You couldn't have imagined where certain ideas would take you. Now looking back, you can piece together stuff, but at the start, you couldn't have charted this path.

One of the CAREER project deliverables is to create a toolkit that other institutions can use to learn from the exemplars. I didn't say this before, but there's five areas that I'm specifically looking at their change strategies of the exemplars, and it's an adaptation of a theory from studying inequity, they're called "inequity regimes" in my workplace. It's related to gender inequities, by realizing that there's inequity in terms of who gets hired, their wages, who gets assigned certain projects. And I think that there's a parallel in higher ed. So I'm looking at admissions, financial aid, the curriculum, I'm looking at chain strategies in some specific categories and developing a toolkit that others can use.

In my copious free time, I am working on a children's book, it's not quite the same, but I've drafted it and the illustrations are done now. It's called "All but the heroes stayed home." It's about the frontline workers of the pandemic.

JW: I definitely want to read it.

3 Initiating

INTRODUCTION

The start of your change project may feel like the start of an epic road trip, the smooth highway rolling out ahead of you with the prospect of unexpected adventures all along the way (see Figure 3.1). Or the start may feel like sitting in a car whose engine is misfiring, full of hesitation and running rough. Whether you come to your change project with your motor already running smooth or if you are not yet up to speed, the contents of this chapter are designed to help you get your project on the road to change. As a result, you will be able to handle the Initiating stage of your project with skill.

For many academic change makers, the early phase of their change projects often consists of similar activities. They act like entrepreneurs, but instead of setting up a business venture, they are creating an educational start up. You may think that referring to your change project as a "start up" is inaccurate, but consider the similarities

FIGURE 3.1 Initiating your change project. (Used with the permission of the author.)

DOI: 10.1201/9781003349037-3

between what experts identify as the early stages of setting up a business and the early stages of the change project you are initiating. Petch 2016, writing for *The Entrepreneur*, identifies the first two stages of a startup lifecycle with terms very similar to what you encountered in Chapter 2 of this book. Stage 1 for Petch is the "soul-searching phase" when you test the feasibility of your idea, much in the same way that you were asked to consider the emerging opportunities for change on your campus and assess your own readiness to take on a change project. Petch writes that Stage 2 in the cycle is launching the start up, and there are many parallels between what you will do in Chapter 3 as you are getting your change project operational.

Even if you have no experience with setting up a new business, you might be able to imagine what tasks would be involved: figuring out how to manage the start up, tracking the work that will be done, finding the right people to do the work, ensuring accountability through a system, setting goals, determining what will count as success, and so on. These tasks are quite similar to what needs to happen as you start your academic change project. You'll need to set goals, identify the right people to work with you, create a shared vision for change, and even confront resistance.

This chapter will present topics relevant to the Initiating stage of your change project, organized by three themes: Communication, Teamwork, and Diagnosing Problems.

- You'll explore the importance of communication for the early stage of your project, particularly the types of communication tools you can use to share the change you wish to promote.
- You'll also consider how you can effectively identify and recruit the right team members and bring them into your change project, using the opportunity to create a shared vision for change.
- And because every change project poses a challenge to "business as usual" in your department and on your campus, you'll look at how you can best anticipate resistance to your change project and understand potential sources of it.

So let's jump into the convertible and crank the engine. You are setting out on a memorable road trip through the landscape of higher education in STEM. You might even want to take a selfie of yourself behind the wheel to commemorate the start!

READ MORE ABOUT IT

Petch, N. 2016. The Five Stages of Your Business Lifecycle: Which Phase Are You In? *Entrepreneur Middle East*. February 29.

THREE COMMUNICATION TOOLS

TOOL TYPE: COMMUNICATION

Summary: In this section, you will use three communication tools to create coherent, persuasive descriptions and stories about your change project.

Peek into any home toolbox and you are likely to find multiple variations on a single type of tool, for example, screwdrivers both straight and Phillips head, hammers that drive nails or tacks. And don't forget wrenches, perhaps the most numerous tool type: a set of fixed crescent wrenches, an adjustable crescent wrench, a set of Allen wrenches (how else will you assemble that new Ikea bookcase?), a pipe wrench, a lug wrench, a monkey wrench, even a can of Liquid Wrench so you can dislodge a rusted bolt. Each wrench addresses a specific task you need to complete, and the best home repair enthusiasts know that they need to have a broad selection of tools if they hope to do a good job.

Communicating about your change project bears many similarities. You may be surprised by the amount of time you spend, not on the work of the change project itself, but communicating about change to your project's stakeholders. As a result, you will need a variety of communication tools at your disposal, each suited to specific audiences and their needs. At the Initiating stage, you need more than one type of communication tool in order to address the different situations, audiences, and purposes you encounter.

In this section, you will learn about three distinct yet adaptable communication tools that will serve you quite well and help you communicate about your project to a variety of stakeholders:

1. Background on Your Change Project: answering the question, where did the idea for your change project come from, and who does it serve?
2. A Definition of Your Change Project: following the template of, what is X?
3. The Change Project Description: answering the question, what is your project about, with details?

BACKGROUND ON YOUR CHANGE PROJECT

As part of your efforts as a change maker, you may be asked, "when did your work on this project begin?," or "what inspired you to devote your time and energy to this project?" You can be a more effective advocate for your change project if you have taken time to consider these questions before you are asked to make a presentation to the Curriculum Committee or at the next campus Town Hall. By developing a background for your change project, you will clarify for yourself why you are planning to do this work and what inspired you to take up the challenge.

Set aside some time to reflect and write responses to these questions:

1. **When did the idea for your change project first emerge?** You might find it helpful to look back at the Emerging Opportunities analysis you did in Chapter 2.

2. **Did the idea for the project come as a result of seeing a particular situation or problem?** You can consider this question from the perspective of individuals who are directly impacted by the situation or problem.

3. **When did you have the "lightbulb moment" regarding this situation or problem?** How did you know that addressing the situation or problem would mean improving conditions for the individuals impacted?

4. **How did you evolve from the initial inspiration to working out your project as the right way to address the situation or problem?** Did you share your insights with colleagues? Did you explore how individuals elsewhere had addressed the same issues?

Once you clarify the background for your change project, you are ready to start to use the rest of the tools in this section to develop communication pieces that can be adapted for various situations and stakeholders. If you need to understand more about how change makers use communication tools like these, check out the Change Maker Interview at the end of this chapter.

A Definition of Your Change Project (A Single Sentence)

Let's start with the most basic tool: a definition of your change project. You need to define what your change project is because it is inevitable that someone will ask you, "what is your change project?" For the purpose of illustration, if you were to ask me, "what is this book about," I can respond with a one-sentence definition:

> *Making Changes in STEM Education: The Change Maker's Toolkit* is a practical workbook based on academic change research and designed to equip STEM faculty with the tools necessary to accomplish their academic change goals.

You can see the form of a definition in several of the sentence's features. Note that it's short. It's a sentence, not a paragraph. It follows the conventions of a definition, the "what is X" formula. And it captures in general terms what the project is. In order to understand what is meant by "tools" and "resources," the one-sentence definition will need to be expanded into a more involved project description.

Why should I spend time, you may ask, writing a short definition? Why not spend your time getting to the longer project description? Simple: writing a definition of your change project is a useful test to determine if your project is clear in your own mind. Writing a definition is also useful to members of your project team, especially if you can use definition writing as an assessment of what each team member believes the project is about. If you have a project team assembled, ask each member of the team to write independently a definition of the project they are working on. Then ask each team member to read their definition out loud to the group. The disparities between definitions will suggest how aligned team members are for the change work they undertake. And this information is valuable early on in your project since you now have an opportunity to orient the team to a shared definition.

Put the Tool to Work

Your definition of the change project:

```



```

Your team members' definition of the change project:

```



```

The Project Description (One Paragraph, About Five Sentences)

Now that you have created the project definition, you can construct the project description in order to expand on the general elements you referred to. Granted, it is a challenge to describe your change project in a paragraph, but limiting yourself to five sentences offers an important lesson in getting to the point regarding your project. In other situations and contexts, individuals in academic fields have ample opportunities to write at length about the change projects they envision. Writing a proposal to the National Science Foundation, for example, will give you roughly 15 pages to discuss your project and how it will impact its stakeholders, although it doesn't seem nearly enough when you are in the midst of writing it. Likewise, writing a research article about your project will provide you with even more pages in which you can provide evidence of your success. But there are few opportunities to capture the entirety of your project in abbreviated form. Of course, you will need to write an abstract for the research article, and the NSF grant proposal requires an abstract that follows a strict format. Despite the apparent similarities, the Project Description is its own type of communication, serving an essential purpose: to distill your project into a single paragraph and provide your reader with a vivid, compelling picture of the project.

At this point, let's go back to the one-sentence definition of this book. Now it's time to add more details in order to flesh out the project description paragraph:

> *Making Changes in STEM Education: The Change Maker's Toolkit* is a practical workbook based on academic change research and designed to equip STEM faculty with the tools necessary to accomplish their academic change goals. [What follows is new material] Each chapter presents tools related to key themes in change work: communication, teamwork, and diagnosing problems. Each tool is also contextualized via essays by the author, herself a change leader in STEM. The author also provides interviews with STEM faculty who are engaged in their own change projects. These interviews offer additional insight into how the tools can be applied to a variety of educational contexts.

Put the Tool to Work

Building on the sentence definition you wrote, it's time to expand into a full paragraph. You may find it helpful to focus on one element of the definition sentence and create a new sentence for it, as is the case in the project description paragraph that expands on the idea of "tools."

Description of the Change Project:

Okay, now let's bank these communication pieces in preparation for the numerous presentations you will make about your change project. One piece of sound advice is to create a slide deck that captures the two pieces—the project definition and the project description—that you wrote. Be sure to give the deck a specific file name, something distinctive, because you are going to go back to the same deck again and again. This is your starting point for connecting with a variety of audiences.

- Have you been invited to talk to your department's Advisory Board about your change project and its potential impacts? Start with your deck and tailor it to the interests of the Board members.
- Will the new dean meet with you next week to talk about the project? Fire up the deck and adapt it to her concerns, like how your project will help increase enrollment and retention of women in STEM.
- Do you have the opportunity to discuss your project with a potential external funder? Check out their funding priorities (usually listed on the funder's webpage) and tailor your description in the slide deck to reflect those priorities.

Whatever adaptations you need to make, your starting point is always the same, with this very important deck.

MY ACTION PLAN FOR THIS TOOL IS:

YOUR DREAM TEAM FOR CHANGE

TOOL TYPE: TEAMWORK

Summary: In this section, you'll consider the best recruits for working with you on your change project.

As an academic change maker, you choose your project team like an effective baseball general manager chooses players. He learns each player's stats: on-base percentage, home runs, walks, hits, and every aspect of their game expressed in numbers. Based on those numbers, he recruits the best team he can afford and embarks on the season, full of hope and a desire to win.

You won't always choose who is on your team for your project. In some cases, however, you have the power to select your team, and for this task, you might be tempted to find your campus buddies and recruit them to the work. Instead, take your cue from baseball and use this important change maker tool: the academic change team player card.

Like the baseball cards you may have collected as a kid, the team player card features, on the front, a solo picture of the team player at their best, wearing the team jersey, smiling, or maybe looking at the camera with dead seriousness, as if to say, "I'm ready to make academic change, put me in the game." Or maybe the photo captures the team player in action, crouching low to scoop up a line drive or swinging through to connect with a fast ball. On the back of the card you'll find all of the important academic change team player statistics: their academic discipline, change teams they've played on, rates of success with previous projects, skills, knowledge, and abilities that contribute to project success. It's all there, everything you need to know about the player in order to make the right choice as you form the team.

And wouldn't it be great if the player card had a few additional stats? After all, selecting the right team players is key to increasing the likelihood of success of the project, as well as the effectiveness of the team members who work together. How about a stat that shows how well the team player works with others, both those on the team and those in the department, college, and campus? Maybe another that tells you how likely the team player is to respond to an email in a timely manner. Definitely, you need an average of how many times the team player listens to, rather than talks over, others in a meeting.

While academic change team player cards would benefit every change maker who is recruiting a team, the cards could benefit the players themselves. In the world of baseball card trading, some player cards hold a higher value than others, a 1909 Honus Waggoner card, for example, or a 1952 Mickey Mantle. Other cards and players fade into obscurity. The value of academic change player cards would reflect their contribution to the project's success and the team's general satisfaction with the work. Consider how an individual's card could increase in value as she achieves her project goals with one team, then is invited to join other projects based on her growing reputation. Her card would be in demand. And which cards would decline in value? The ones for players who can't work collaboratively on a team, who put their own goals ahead of the team's, a player about whom every other team member declares, "Never again!"

It's unlikely that you'll see a line of academic change player cards issued from Topp's any time soon, but that doesn't mean you can't prototype the concept by identifying the skills and traits you seek in prospective team members. But before you set off to the Hot Stove League and start trading for team members, take a moment to reflect on what you need in the team members you seek.

PUT THE TOOL TO WORK

Like a skilled general manager for a successful team, you need to identify the knowledge, skills, and abilities (KSAs) required on your team in order to move the project forward. In order to approach team recruitment in the most rational way possible, you'll start with project goals and the KSAs required to accomplish the goals (Pentland 2012).

Using the grid below, list each project goal, then inventory the KSAs you believe you'll need in order to accomplish the goal. For example, your change project may relate to establishing a community-based project that requires your department to collaborate actively with community members. As a result, your team is probably in need of skills and knowledge that include the ability to cross the town/gown divide, the ability to listen carefully to the needs that the community (not the college) defines, and familiarity with the local environment and context.

Project Goal	Knowledge, Skills, Abilities Required	Possible Team Member(s)
Project Goal		
Project Goal		
Add additional goals as needed		

In order to recruit the right team members for your project, you'll need to take time to identify individuals who can provide the necessary KSAs. Start with careful observation, since you are scouting out possible players for your team (see, the baseball metaphor continues to be apt!). And don't neglect to consult colleagues in order to expand the reach of your recruitment. You may realize, however, that you need a specific skill set on the team, but you don't personally know anyone who fits the bill. This is an opportunity to scan your institution's departmental webpages. Does your college or university have a departmental devoted to this area of knowledge? If so, then it would be advantageous to attend a talk by one of these department members, or search out their published papers. Think of yourself as a major league talent scout who is on the lookout for a new prospect.

The key here is to make your selection of personnel more a matter of talent identification than proximity. Likewise, you may have had great experiences working with your favorite colleague in the office next to yours, but the skills you actually need are located across campus. This is the time to set aside an emotional choice and select instead with your rational mind. Your team, and your project, will be all the better for it.

Read More About It

You can find quite a lot of research on team selection and formation. If you aren't ready to tackle it all, you can start with Sandy Pentland's article. It offers a research-based approach to choosing the right team members for your project.

Pentland, S. 2012. The new science of building great teams. *Harvard Business Review*, April 1.

My action plan for this tool is:

THE WORST TEAM IN THE WORLD

TOOL TYPE: TEAMWORK

Summary: Many change makers have had negative team experiences. By assessing team members and their preferences for working on a team, you can define expectations you have for the team and the expectations they have for each other.

When faculty initiate a team project in their undergraduate classes, they may begin asking their students to reflect on an important topic: describe your best and your worst team experiences with a class project. Students' responses are not surprising. The best team experiences, they often write, feature each team member performing their tasks well and on time. Students describe clear communication without conflict, as well as the team's ability to use each member's strengths to great advantage. While such reflections are wonderful to read, the most useful responses are the ones that identify what went wrong on the team. These narratives often feature a team member who isn't willing to contribute at the same level as the others. This team member may check out at some point in the project, leaving the other team members to pick up the slack. This team member may be willing to work on a specific aspect of the project but then not possess the necessary technical knowledge to complete it. This team member may be unwilling to listen to the input from other members and refuse to pursue any direction for the project that isn't solely his own. With this information, the professor can design teams with members who can more effectively complete the assigned tasks. Unfortunately, even with such planning and care, student teams sometimes devolve into the Worst Team in the World.

It is tempting to believe that inexperience and youth are the root causes of undergraduate teams' failures. As faculty, you may envision that your students will graduate, pursue their professional careers, and never encounter problematic teams ever again. Every professional, every academic, knows better, based on experience. If you were to complete the same reflection assignment about committees you have served on, there is a good chance that you might describe a best and worst team experience in much the same way. Of course, some students believe that all of their troubles could be avoided if they were only allowed to pick their own team members for the class project. The irony of the best and worst team reflection assignment is that choosing their own team members doesn't necessarily predict a better or worse result. Picking their own teams for the group project won't prevent some teams from getting into trouble. A group of friends, for example, decides to collaborate for the purpose of the class project. Often this works well, but sometimes it fails completely. In this case, however, the only persons the team can blame are each other.

Earlier in this chapter, you learned a strategy to help you select members of your team based on the goals of the project and the knowledge, skills, and abilities you require to best meet those goals. If this is your situation, then conducting a team assessment is important. The assessment should include asking each team member to describe their habits, their communication preferences, and other aspects of their way of completing tasks that can help foreground differences in how they approach their work. In contrast, as is the case with undergraduate team projects, you may not have the choice of who will be on your team for your change project. You may be charged to create change through the work of a standing committee or a pre-established group. One source of team dysfunction is incorrect expectations on the part of team members. The same problem with expectations lies potentially in all teams, and it is best to start every team project with some time to inventory of the preferred approaches of each team member.

It is common practice in companies and corporations to subject employees to a version of a personality test like the MBTI, and you may perhaps wish to initiate your team's work with a version of that tool. It is more important to the success of the team, however, to understand what each team member's preferences are, since these define how the individual team members will communicate and collaborate. Once you have collected responses from each team member, you can bring the team together to review results and design a work flow that takes the differences into account.

Put the Tool to Work

There are several different team assessment tools you can use for this purpose, and one of these is Dr. Joanne Wolfe's powerful tool that helps team leaders and team members set clear expectations and preferences before a team project begins (Wolfe 2010). The Team Norms Self-Assessment, while a useful way to probe team members' preferences and expectations, should be deployed carefully and with the caveat that any assessment tool can be mishandled if not done with good intentions. The purpose is to bring forward issues that team members may be less willing to talk about at the start of a project but may rise up and impact the team's work later on. Just remember that the point is to focus on preferences, not on pigeon holes!

The ideal time for you and your team members to complete the assessment is before the team project begins, but it can be introduced at any point in the project if the team encounters conflicts resulting from different, unshared expectations. Once each team member has completed the assessment independently, the team can discuss the results together and use the evaluation to set clear expectations for working together. The information can also provide the basis for the Collaborators' Compact, which is discussed in the next section.

Read More About It

Wolfe, J. 2010. *Team Writing: A Guide to Working in Groups*. New York: Bedford/St. Martin's Press.

My action plan for this tool is:

COLLABORATORS' COMPACT

TOOL TYPE: TEAMWORK

Summary: Building on the assessment of team members' preferences, the Collaborators' Compact helps teams create shared expectations for their work on the change project.

In the first two sections related to Teamwork in this chapter, a change maker can use specific tools to find the right team members, get to know them, and clarify what each of them needs in order to do their best work. The third useful tool at the Initiating stage addresses the finer points of collaboration between team members. When disciplinary experts meet in order to collaborate, they may not share a common language since each is immersed in the vocabulary and knowledge framework that characterizes their field of study. When the biologist and the literature professor begin their work together, they may not sufficiently understand that the biologist conceptualizes the world and her work in terms of systems, while the literature professor conceptualizes the world in terms of metaphors and interpretation. Likewise, the lack of a common language can mean that team members talk at cross-purposes, not realizing that they may need to translate between the fields (Stone et al. 2010).

Equally important are the differences between collaborators and their habits of work and expectations for the project. One team member may thrive on starting their day early, responding to emails within 30 minutes of receiving them, and responding to document drafts quickly and copiously. Another team member may work best late into the night, scanning but not responding to emails unless assigned a specific task to complete, and preferring to provide holistic feedback. Both approaches are correct for the specific team members, but expectations and assumptions regarding how team members should work are the source of quite a lot of consternation and conflict. The purpose of the Collaborators' Compact is to head off these avoidable conflicts at the start of the project, through a purposeful analysis of what makes each team member unique while building bridges between the disciplines and their intellectual frameworks (Schein 2013).

As the person who is recruiting the team and/or leading it, you have pre-work to do before you use the featured tool in this section: the Collaborators' Compact. Through a series of challenging questions, you can reflect on your role both as the person who is assembling the team and as a member of the team.

Question 1: Why are faculty and academic staff coming together to start a collaboration? Is there urgency for this collaboration?

This question may seem to be the provenance of the team convener only, but consider how this question may strike a team member, rather than the team leader. It can spark reflection on their part regarding the nature of the project and the urgency to get the project done. While the team leader may feel the urgency keenly, that sense of urgency may not be shared by other team members. The potential conflict arises when the drive of the leader doesn't match the drive, or lack of it, among the team members.

Question 2: The individuals who must work together may come from different disciplines. How does each discipline define its primary problems? What frames does each discipline use to formulate its approaches? How can we bridge those differences?

Each team member is recruited for the team in order to offer specific contributions, usually from the perspective of their discipline. Disciplinary differences, however, pose a unique challenge for teams, since much of the disciplinary framework and vocabulary remains unacknowledged when the team comes together in its early stages. For example, the customary digging into theoretical foundations that characterizes the field of philosophy may come into direct conflict with the practical, applied perspective that forms the field of software engineering. And without some discussion of how these differences may lead to conflict, the team is potentially headed toward trouble.

Question 3: How can the team cultivate interdisciplinary and transdisciplinary collaborations through effective communication? What communication tools and methods offer the most promise in these contexts?

The remedy may be obvious. The way to cultivate collaborations across disciplines is to encourage conversations between team members and to highlight the moments when the frameworks clash against each other. This is where the Collaborator's Compact comes in. The purpose of this tool is to help you identify the elements of any collaboration that are most important to you and to encourage you to share these expectations with future collaborators.

PUT THE TOOL TO WORK

Once you have collected your own thoughts about collaboration in the pre-work, it's time to bring your team together for their own pre-work in advance of creating their Compact. Consider asking each team member to work on these questions independently, followed by a group discussion of individual responses.

For the individual team member:

Consider a previous team collaboration and list the positive aspects that you experienced in the work. For example, you may have worked with a collaborator who shared your work habits, like grading student assignments within 48 hours.

Again for the team member, reflecting individually:

Every discipline operates from a framework of assumptions that are second nature to the faculty who are members of that discipline. Use the space below to state what a few of those assumptions are in your field. For example, in my field of literary analysis, I work with the assumption that multiple interpretations of a text are possible, but that doesn't mean that every interpretation is equally valid.

Now for you, as the team's convener:

Take a moment to review the Collaborators' Compact document template below. With your own assumptions and expectations that your team members reflected on above, your next step is to encourage such a metaphorical "putting cards on the table" from the individuals you wish to collaborate with. Their responses, and the ensuing discussion, are the data from which you can create your own version of the Collaborator's Compact. The template is a starting point for your discussion with your collaborators, and you should feel free to revise and adapt the Compact to your specific project and campus context.

The Collaborators' Compact [list the names of those who are collaborating]: We are entering into this collaboration in order to achieve/produce:

- [outcome 1]
- [outcome 2]
- [outcome 3]

We believe impact of this collaboration will be:

- [on stakeholders]
- [on ourselves]
- [on our fields of expertise]

After discussing the disciplinary frameworks of our different fields, we now agree to operate from the following set of shared assumptions:

- [assumption 1]
- [assumption 2]

We intend to adopt the following work habits as much as possible:

- [grading]
- [meetings]
- [communication]

If we discover that our collaboration is producing conflict and disagreement, we agree to employ the following strategies:

- [communication strategies]
- [neutral third party]

We will revisit this compact at intervals during the collaboration:

- [mid-term]
- [final week of term]
- [three/six/nine months later]

READ MORE ABOUT IT

In addition to the book *Difficult Conversations* (Stone et al. 2010), you and your team may benefit from having additional resources available, even perhaps as a reading list for the team. In that spirit, consider the *Harvard Business Review* resource series that includes *HBR's 10 Must Reads: On Communication* (2013), as well as Edgar Schein's excellent book *Humble*

Inquiry: The Gentle Art of Asking Instead of Telling (2013). You can provide good support for the work of your team by reading selections together and discussing them as a change community of practice.

HBR's 10 Must Reads: On Communication. 2013. Boston, MA: Harvard Business School Publishing Corporation.

Schein, E. 2013. *Humble Inquiry: The Gentle Art of Asking Instead of Telling*. Oakland, CA: Berrett-Koehler Publishers.

Stone, D., B. Patton, and S. Heen. 2010. *Difficult Conversations: How to Discuss What Matters Most*. New York, NY: Penguin Books.

MY ACTION PLAN FOR THIS TOOL IS:

CREATING AND COMMUNICATING A SHARED VISION FOR CHANGE

TOOL TYPE: COMMUNICATION AND TEAMWORK

Summary: A change project requires a shared vision that is created collaboratively by the change makers and communicated to stakeholders.

At first glance, there may not seem to be many similarities between the fields of literary interpretation and academic change making, but each change project can be viewed from the perspective of multiple individuals who have a stake in it. Creating and communicating a shared vision should include a range of interpretations of exactly what the change project seeks to achieve.

Unfortunately, most change projects start as the vision of one person or a small team and stay that way through to the project's conclusion, usually because the project vision is formulated as part of an initial proposal. Proposal writing is a solitary effort, or the work of a small group. Once the proposal is funded, however, the work of the project expands out from the original core and includes faculty, staff, administrators, students, all focused on the vision that was devised for the original proposal. But why would an individual who was not engaged with the formulation of the original vision of the project dedicate their time and effort? What stake do they have in the project as it was envisioned by others? A project vision usually emerges from a two-stage process in which the vision is created (usually by project leaders without stakeholder engagement), then followed by sharing the stated vision (but without taking stakeholder feedback, revising the vision, or adapting the vision to reflect their needs). In this chapter, you will learn a different view of creating and communicating a vision, a view that reflects a dynamic, ongoing process that runs through the life of the entire project, not just in its beginning stages.

The source for this alternative vision work emerged from the Revolutionizing Engineering Departments (RED) project funded by the National Science Foundation beginning in 2015. When six RED teams were funded in 2015 to transform engineering and computer science education, the RED Participatory Action Research (REDPAR) project was also funded. REDPAR is a collaboration between Rose-Hulman Institute of Technology and the University of Washington to support the RED change makers through a customized change curriculum while studying and conducting research in the processes of academic change. The REDPAR project team worked with RED team members each year, and in the process, it became clear that the most effective, successful teams took time at the start of their projects to define a vision for change that brought all team members in. The RED projects are, by their very nature, multidisciplinary, and team members are often not co-located. When a team engaged substantially with the vision for their project, taking time to develop a shared vision for change, they experienced less internal and external resistance for their projects. Their communication work was also greatly facilitated by the process of creating the vision (Doten-Snitker et al. 2020).

As part of a paper on shared vision in academic change projects like RED, the REDPAR team made use of several important concepts from the change research, including the principle of "co-orientation." Co-orientation (Taylor 2016) invites stakeholders into the visioning process and allows project leaders and stakeholders to tune into each other and commit to the project. The tool below is meant to give you and your team a structure by which you can co-orient and tune in to each other and to stakeholders, through the process of creating a shared vision for your project.

PUT THE TOOL TO WORK

Creating the shared vision can best be accomplished through a multistage process with individuals defining what they believe the vision of the change project should be, and then working together to create a vision that reflects the views of all team members.

Working Individually: Independently and in a couple of sentences, each team member should describe the change project. Individuals should write the description without consulting others or with the project or team leader.

Description of the Change Project:

```
┌─────────────────────────────────────────────────────────────────────┐
│                                                                       │
│                                                                       │
│                                                                       │
│                                                                       │
│                                                                       │
└─────────────────────────────────────────────────────────────────────┘
```

Next, each team member on their own should write, in one short phrase, the ultimate goal of the change project. What is the intended outcome, what is the team trying to accomplish, what need is being met, what problem is being solved?

Outcome(s) the team is trying to accomplish

```
┌─────────────────────────────────────────────────────────────────────┐
│                                                                       │
│                                                                       │
│                                                                       │
│                                                                       │
│                                                                       │
└─────────────────────────────────────────────────────────────────────┘
```

Together as a Team: Now that each team member has had a chance to reflect on the change project and define their personal conception of it, it's time for individuals to share their own responses and compare them to those of others. The process then is to derive the shared vision for the project incorporating the individual visions as much as possible. It is important here that each team member feels heard and acknowledged for their contribution to this vision. By doing so, individuals can develop a firm stake in the project, since some dimension of the vision connects directly to them.

Draft the Shared Vision, based on the vision of individual team members:

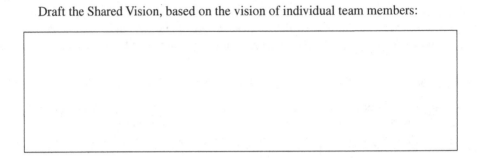

Making Time to Cultivate Vision: Going forward, you should consider scheduling periodic revisiting of the shared vision, both among members of the team and with stakeholders as the project moves forward. The vision may need to be adapted, revised, or otherwise reformulated if it is to continue to engage team members, stakeholders, and others who are engaged with it. And if the project seems to be running off the rails, a return to the shared vision may be a great way to reconnect everyone involved.

READ MORE ABOUT IT

In my collaboration with colleagues serving on NSF RED teams, we have identified a number of important insights about how change teams start their work and how they can be successful through the creation of shared vision. For the RED program, we have designed a Start Up Session that helps new teams identify their individual and shared goals for the work ahead (Williams et al. 2020). In addition, we have published our research findings regarding the creation of shared vision as a necessary early stage for change teams (Doten-Snitker et al. 2020). These resources can provide additional support for the development of your change team.

Doten-Snitker, K., C. Margherio, E. Litzler, J. Williams, and E. Ingram. 2020. Developing a shared vision for change: Moving toward inclusive empowerment, *Research in Higher Education* 62: 206–229. DOI: 10.1007/s11162-020-09594-9

Taylor, J.R. 2016. "Co-orientation." *The International Encyclopedia of Communication Theory and Philosophy*. Wiley Online Library. DOI: 10.1002/9781118766804.wbiect003

Williams, J. M., S. Mohan, E. Andrijcic, C. Margherio, E. Litzler, and K. Doten-Snitker. (2020, June), *The RED Teams Start-Up Session: Leveraging Research with Practice for Success in Academic Change*. Paper presented at *2020 ASEE Virtual Annual Conference Content Access*, Virtual Online. DOI: 10.18260/1-2--35360.

MY ACTION PLAN FOR THIS TOOL IS:

WHEN CHANGE MEANS LOSS

TOOL TYPE: DIAGNOSING PROBLEMS

Summary: This section addresses one possible source of resistance to a change project and offers an empathic approach to naysayers.

Writing for the Mind/Shift website (where the focus is on K-12 education), Katrina Schwartz takes up the issue of how school principals and administrators can help teachers and staff deal with change (2017). In summarizing the work of psychologist Robert Evans, Schwartz suggests that change for many people isn't about "growth or capacity-building or learning; it's about loss." Evans has written extensively about change, but his observations are somewhat surprising: "Resistance to change is normal and necessary . . . If you are part of some big change in your school and you aren't expecting resistance, there's something wrong with your plan" (Evans 2011).

How often do change agents expect a positive, enthusiastic reaction to their plans for change? Working on their own or with a team, they devise a new approach, program, or initiative and expect others to be as excited about it as they are. Then they are surprised when they encounter initial resistance. The change makers may dismiss the concerns of others with phrases like "They just don't get it!," or "They hate change!," or "They will appreciate all of our hard work once they see the change in action!" The tendency of many change agents is to drive on, discounting this resistance, maintaining focus on the change itself, and disregarding the associated interpersonal issues that surround it. In the end, the change might be in place, but the price paid by the change agents can impact any future effort.

So how can change agents approach resistance differently, before even launching the project? What if they take time to reconceptualize resistance as loss and then empathize with those who are impacted by the change? Schwartz notes:

> It's rare for anyone's first reaction to a call for change to be all positive. Much more often those pushing for change don't realize that they are devaluing everything colleagues hold dear.
>
> Sometimes the call for change makes people feel like everything they've been doing up to that point has been wrong and bad for students. Worse, it can sound like a devaluation of how the teacher learned and, by extension, those who taught her. That's a personal loss. Educators react negatively when they are asked to change not because they don't want to do what's best for kids, but because they feel bereaved.
>
> (2017)

In higher education, the approach to change could be radically different if change makers acknowledged the loss that their change projects could potentially mean for faculty, staff, and administrators. In addressing the loss, the communication messages might sound remarkably like condolences offered to mourners: "I know that this change affects you and your work significantly" or "By making this change, we are altering what is familiar and important to you." You may have never seen this approach used by members of the administration at your own institution, but it is possible that such a realignment could shift the perspectives of faculty and staff.

Referring to the loss and acknowledging its impact on the individual does not alter the reality of the change itself. There may be an opportunity, however, to bring faculty and staff into a process of shaping plans for the future into a revised, shared vision. One possible response to loss is building something new, but change agents often ignore or exclude others who could be impacted by the change, delivering the future plan as a fait accompli. Consider coupling the acknowledgment of loss with a concrete effort to craft a shared vision in partnership with those who are experiencing the loss. By bringing these new partners into your effort, you create a path forward toward the change you and your collaborators can share together.

Put the Tool to Work

It may be difficult for you to express empathy and understanding for those who resist your change. When confronted with a colleague who resembles Herman Melville's character Bartelby the Scrivener (and his repeated phrase, "I would prefer not to"), you may feel unprepared to respond constructively to the assertion. In order to feel prepared for the next encounter with resistors, let's take a moment to catalog the objections you have heard thus far to your proposed change project. Once you have a list of objections you have already heard, use the space on the right side of the page to write an empathetic response that reflects your acknowledgment that change may mean a loss for this person. If you find it difficult to generate appropriate empathy, try to view the objection as if it has been leveled at your current change project but ten years into the future, once it has been established and integrated into your campus culture. How would your future self react if an up and come-er in your department proposed a change that would make your initiative obsolete? By adopting empathy, you may discover that there is much more to appreciate about this Bartelby than just their resistance.

Objection I have heard about my project	A possible response that would acknowledge this change as loss
1.	
2.	
3.	
4.	
5.	

Read More About It

Evans, R. 2011. The Human Side of School Change: Reform, Resistance, and the Real-Life Problems of Innovation. San Francisco, CA: Jossey-Bass.
Schwartz, K. 2017. "How School Leaders Can Attend to the Emotional Side of Change" https://www.kqed.org/mindshift/49486/how-school-leaders-can-attend-to-the-emotional-side-of-change

My action plan for this tool is:

FLYING MONKEYS

TOOL TYPE: DIAGNOSING PROBLEMS

Summary: In this section, you confront your anxieties about your change project in its early stages by identifying potential problems and assessing the likelihood of the problems emerging.

For children, the flying monkeys are terrifying. Every year, the *Wizard of Oz* is broadcast, and the same scene sends kids running to bury their faces into the nearest sofa cushions. The Wicked Witch of the West commands her flying monkey henchmen to capture Dorothy and bring her back to the castle. So off they fly, straight to their target. Dorothy and her friends see them coming, a flying squadron announced by barks and flapping wings. The quartet doesn't stand a chance. The Tinman wields his ax but is quickly disarmed. The attackers disembowel the Scarecrow, pulling out his straw stuffing. Despite his threatening teeth, the Lion can't connect with flying monkey flesh. Instead, the monkeys grab Dorothy, lift her up, and take off for what seems to be her doom. And she screams all the way.

Sounds are key in the scene. Even before they make their entrance, the monkeys can be heard: their flapping wings, their calls like barking dogs. Yes, the sight of them is disturbing, but the sound is somehow worse. Taken in the context of the entire film, the horrific kidnapping of Dorothy is a turning point. Following Toto's lead (who knew that a terrier could track someone from the ground who was transported by air?!), the three companions locate the castle and plan Dorothy's rescue. And don't forget that Dorothy is the one who finally does the old witch in with one inadvertent toss of the water bucket. So long, Wicked Witch of the West!

Change makers may not harbor many irrational fears, but most do experience anxiety about change projects that may go awry. It's the catalog of "What If's" that give change makers sleepless nights and waking dread. What if your great idea isn't accepted or valued by someone with power and control over the project? What if poor outcomes cause the entire effort to collapse in on itself? Unfortunately, focusing on the "What If's" can often keep change makers from enjoying and valuing what is going right in the project.

You can call these fears and frights Flying Monkeys, or you may prefer to call them by another name that is more descriptive for you. Flying Monkeys pose real dangers and threats to change projects, and the sight and sound of their approach are terrifying. But are they really as bad as your fears paint them to be? If you turn and confront them, can you subdue them? Could they actually propel your future actions? Can you toss cold water on them and rob them of their power? Let's call them out!

PUT THE TOOL TO WORK

Brainstorm—Make a list of everything that could go wrong with your change project. Put your pen to paper and continue to write without lifting up the pen or removing your fingers from the keyboard. Don't judge or edit, keep writing.

```

```

Using your brainstorming list, map each Flying Monkey into a vertical continuum according to likelihood: "Certain to happen," "Likely to happen," "Not very likely to happen," "Not happening." We don't need to waste time on monkeys that squat at the bottom of the list. Instead, pick one or two monkeys that are hopping and barking near the top. Take a moment to propose two actions or strategies that could rob the monkey of its power. Think concretely and use your power to take these monkeys down!

MY ACTION PLAN FOR THIS TOOL IS:

LET ME GIVE YOU SOME BAD ADVICE

TOOL TYPE: DIAGNOSING PROBLEMS

Summary: To conclude Chapter 3 Initiating, consider what it means to start a change project, even if you are early in your academic career.

As a new faculty member, you may receive helpful advice offered by well-meaning colleagues who wish to remove obstacles from your professional path. One suggestion often overhead: don't speak at faculty meetings during the early years of your career. This is, perhaps, the worst advice provided to untenured faculty. By instructing colleagues that they should keep their opinions to themselves until they are assured of their tenure, we are guaranteeing that they will be silent even after they achieve their professional goals. Once you have lost the habit of speaking up, it's much harder to relearn it.

For change makers, the advice is particularly bad. If you are reading this book and working on a change project, there is no way that you can silence yourself. Unless you speak, how are you going to advance the project? How will you recruit individuals to join the project team? How will anyone know what you wish to change and why and what benefits you believe will accrue when the project is implemented? So let me give you some necessary advice: don't keep quiet.

Don't keep quiet, but also consider this caveat. Before you speak, take the time to ask questions and listen. This advice is particularly relevant when you reflect on the obstacles you face in moving your change project forward, obstacles that come in the form of colleagues. It is tempting to scan the faces at your department meeting and assume that you know beforehand who will support and who will resist your proposed change. You don't always know, however, who is potentially in your corner so it is best if you find out who the obstacles are rather than assume you already know.

Yes, explicit challenges can be heard and seen, but don't assume that you understand the speaker's motivation:

- Are they objecting just to object, or do they have a reasoned stand?
- Can you engage them in a productive way?
- Can they be convinced to give your change a try?

So much of the objecting stance has to do with personal motivations, rather than rational arguments to the contrary. Consider your colleagues from a perspective of a distant past, when you were not a member of the department, and they were the freshly minted PhD who was hired by the department. It is possible that they were once the change maker, full of new ideas, eager to embark on transforming the curriculum, the department, the college, the world! If all you see now is the old guard faculty member who stands in your way, reframe your perspective and consider what they might have appeared and sounded like 20 years ago. Is there a historical picture of the faculty member in the university archives that shows this faculty member in their prime? Dig that out and consider it as their true image. Now you can build a portrait of them as the change maker in their prime, and you are their rightful heir for change in the department.

It's important to remember that many people choose an academic career because it is (or used to be) set out in clear stages with defined outcomes: dissertation, first job, promotion to tenured associate professor, and promotion to full professor. For those who seek out stability and consistency reflected in the traditional academic path, ambiguity will feel like a threat. Change making will not be what they choose because it threatens loss of control. These colleagues will probably object, either vocally at a department meeting or behind your back when you aren't in the room, to the change you propose to make. Your graduate advisor may well say, "well, I'd recommend that you don't highlight your change aspirations while you are on the job market." Your senior colleagues might recommend, it would be best to keep that idea to yourself until you get tenure. You may hear that it would be better to avoid controversy and keep silent during your early years in the department. Just remember that by taking what others consider good advice, you risk losing interest in your project; you may lose the energy, perhaps the opportunity, to make the change that you believe is important to make. What was important to do early in your career may seem less so later on.

But that doesn't mean ignoring the realities of career advancement and what it takes to achieve your professional goal. The point is to learn the tools that can help you pursue your change making goal with a greater likelihood of success. Apply the same attention to your change making skills development that you did with your research and teaching. In both of those areas, you had to develop skills, acquire knowledge, and develop abilities, if you were going to complete the requirements for your degree. And just because you were never taught the change maker tools as a student doesn't mean you can't acquire them now. Take a look around you and you'll notice that there are plenty of change makers in your general vicinity, people in all roles and jobs who you can look upon as role models for the work you want to do. Watch them carefully and learn from them. That's pretty good advice.

My action plan for this tool is:

CHANGE MAKER INTERVIEW: DR. RACHEL MCCORD ELLESTAD

By way of illustrating the change maker tools you acquired in this chapter, I'd like to introduce to a change maker who has experience with putting a change project into motion, even though she is in the early years of her career. During the interview, Dr. Rachel Ellestad expressed her view of why she is a change maker:

> I have a lot of friends who say, "Oh, I don't like conflict. I don't like it when we have disagreements." I look at that situation differently. I think the only way that you can make progress is if you identify where you agree and where you don't agree, and then you try to work through that. This reflects a core idea about change. You should never be happy with the way things are. Things can always be better. So, the idea of change doesn't bother me. I think it's just so natural.

Dr. Ellestad is the Director of Engineering Fundamentals in the Tickle College of Engineering at the University of Tennessee in Knoxville. She holds a BS and MS in Mechanical Engineering and an MBA from UT Knoxville. She also holds a PhD in Engineering Education from Virginia Tech. She worked as a Manufacturing Technology Engineer with DuPont Chemical Company in Richmond, VA, in the Nomex® Fiber Spinning Process. She also worked as a co-op engineer for DuPont in Parkersburg, WV, and as an engineering intern with RTI Fabrication in Houston, TX, in the titanium fabrication facility. Both her educational and work experiences inform Rachel's approach to creating an educational environment that both challenges students to take ownership of their learning as well as shows students the practical side of what they are learning, in their coursework. As Director, Rachel leads a staff of 12 full-time faculty, 2 full-time staff, and over 60 student workers to support the over 1,000 students that join the Engage Engineering Fundamentals Program each fall.

In talking to Rachel, I was particularly impressed with how she applies her training as an engineer to the nontechnical aspects of her work as a change maker. She illustrates that aspects of your disciplinary training have an application to the change making endeavor.

JW: Rachel, you were recently named Director of the Engineering Fundamentals program at the University of Tennessee Knoxville College of Engineering, and in that role, you have a number of different change projects going simultaneously. One of these is the creation of a department of engineering education at UT, so I'm excited to think about what that future could be. And you are doing this work in a period of leadership transitions at your institution. For example, the dean of the college is an interim, but you're not waiting for the permanent dean to be chosen. You're pushing ahead.

RME: We're trying to as much as possible. As a person that's in an interim position, when you're working with a person that's in an interim position, that person tends to be only willing to do so much, knowing that either they may be stepping into the full-time role or someone else may be stepping into the full-time role. So the idea of setting a vision and moving forward with it when you're in an interim role can be challenging. So that's the position that we're in. The interim dean wants to be as proactive as possible, but he also realizes he's the interim in the position and a decision like moving forward with the engineering education department or moving forward with some sort of new organization probably isn't the best thing for an interim to do and is best left to person chosen to be the dean.

So, we're kind of in limbo there, but I'm trying to take the opportunity to conduct the education again of the administrative staff as much as possible, because it's been an education process every step of the way, trying to educate people that are in decision-making roles about what the engineering education discipline is, what it can do for a college, and how it's an opportunity for the College of Engineering to engage in a new type of organization, how it really can benefit the college and the university as well. But you have to educate, and then if there's turnover, you have to educate again and then turn over and educate again. So it's a little bit of an iterative process until you can actually move forward.

JW: So as you educate and educate again and educate again, you've probably become quite skilled at doing the education in a way that takes hold with each person you need to educate. Can you tell me a story about educating a dean or other administrator?

RME: We started the talks of this process five deans ago and two sets of associate deans ago. Like many other institutions, we've had turnover at the associate dean level and the dean level. Initially, when we were trying to educate, we were, and I say, "we" because it's another colleague and I who are the first people of this discipline here. We went into initial meetings and what we posed to them was what are our needs as faculty members would be in a new department. We said that we would like to be able to do this kind of stuff and just so you know, here are some needs that we have. We thought that that would be good for them to know and I'm sure it was good for them to know, but it really wasn't what their interest was.

Now it seems very obvious that this is what we should have done, but while it's important to present your own needs, you've got to be able to communicate how you can meet someone else's needs as well and do so very quickly. We learned that very quickly. My colleague and I learned that some of the language that would help communicate our benefit to the college would be things like how many programs like this are out there and where are these programs located, at what type of institutions, what type of research funding is available, what types of awards, what types of notoriety do these programs get that can bring some notoriety to the college, how many NSF Career Awards have people in the discipline won.

All of these things speak to the measurable metrics that administrators have to think about, and that's what you have to think about when you're trying to move a project forward. I don't think that's something that as an individual faculty member I would have thought about because I would have thought about the things that I need to do my job, and I know that it's important, so it should just be important to everybody else. But you have to be able to communicate importance on a level that makes sense to a lot of different functions.

We had to communicate to people. We had to figure out how to package what the benefits are to the college by having this type of department here. Eventually we developed a good, tight story, so that every time we had a new dean come in or when we had our new associate dean come in who I now report to, we had to start over with her, but we had this presentation that described how these programs are organized, here are the different ways it has been done, here are the types of schools that do the different organizations, here are the number of grants they bring in and the number of faculty that they have in each unit and how many PhD students they put through. We already had that. We didn't have to go back and do that work again. So we can almost anticipate some of her questions going in, and we were able to do that. So the education process got a little bit quicker every time because we were prepared for that.

JW: So with five deans, there were five opportunities to refine your story and make your work completely airtight.

RME: Absolutely because every time you have to go through that process again, somebody may ask one or two new questions that make you think about something new, something that is important for us to think about. So it's probably good that now we're ready for it. Now we have this whole slide deck where we've just added things to it and when it came time to start thinking about actual formation of the department, they said, "Oh, we've got all these questions we need answered." And you know, my colleague and I looked at the sheet and we went, "Okay, we got all this. We're ready. Here we go." So yes, I would completely agree it definitely helped refine the story as we went through the process.

JW: I think you're being encouraged to keep working with these deans, associate deans and getting some encouragement, positive response to keep going. Are you getting that kind of positive response from faculty as well?

RME: Yes and no. So if I'm just honest, it depends on who you talk to and what their goals and objectives are in the college. So we have a number of individuals in the college who are really receptive to the idea of having some sort of organization, again, whether it's a department or whether it's organized in some other way, and who are supportive of the idea moving forward. Some are faculty, some are department heads, some are staff. They definitely see a benefit to the college for us having some different organization. And then there are definitely some people who are not interested. They either don't see the value of engineering education as a discipline, so they just don't see it as necessary. I think it's reasonable for people to look at it this way, because they see it as a competitor for resources. I can't argue with that. I can't argue with the fact that bringing in something new has the potential to divert resources away from people and programs that are already there.

It's a valid concern, so if we want to organize and move forward with having something in the college, we have to be able to work through that type of barrier and either say, yes, there is some competition for resources, but here's our ability to generate and so we will actually generate an offset to what we have to compete for. Or two, yes, we're going to compete and let's go at it. We're going to compete with you and if you're scared, then maybe you should pick up your game a little bit because that's part of operating in this space.

JW: So you're not ruling out the fact that you may just take those naysayers on at some point, but you don't have to do that right now.

RME: Right, we don't have to do that right now. A lot of the work is getting over hurdles and barriers at this point and at that level is above me. Getting department heads on board with an idea, I can provide information and resources to a higher administrator who can then work on gathering that support and those people. Or the administrator can just tell them, "We're doing this." It could be that they work to get everybody on board. They work to get as many people on board and then they tell other people we're doing it or they don't work to get people on board and we don't do it. There's a lot of different paths that could be followed.

JW: I see how you are directing your communication effort up to levels above you. What about directing it to faculty who are joining the UT College of Engineering on the tenure track? How has your communication worked with them? Because I assume that you've been approaching them as well.

RME: We have some collaborations going on now, five faculty who are trained in this area and have research in pedagogical collaboration among the college. We have a number of people who have been working in different ways to have collaborations in the different departments. For example, a faculty member in our unit who does a numerical methods class, he's trying to do innovative pedagogical work and he's trying to employ the departments to give him real world problems and real world data sets to bring into his classroom and have that collaboration and connection. So that's one really positive way that he's been reaching out to almost every department in the college.

We have several other faculty members including myself who have been on grants with other departments. We have sought funded grants with our material science department and our electrical engineering department to have a collaboration with them. We're trying to build networks of support, but also just networks of people to work with in that capacity. We've also been working with people outside the college too, with the College of Education. We also have partners in the College of Communication where we've done educational work. We actually have a biology education faculty member in arts and sciences and so we have worked with her too. So we have a number of cross collaborations in different colleges as well to garner support and just collaboration at the university level as well.

JW: How did you pick those particular people as partners and approach them? What was your rationale there?

RME: We tried to participate in educationally minded activities outside of the college, and we just got to know people who were doing cool things. We said, "Hey, that's cool. Can we work with you on that?" Honestly, that's the type of collaboration that we would do in the engineering ed community too, where we would listen to somebody give a presentation and think, "That's really neat. I think that lines up with something I could do. Let's find something to work on together." So we've tried to do that out in the university community as well. Sometimes people have come to us too, especially since funders like the National Science Foundation require educational research components. Faculty members think, "Oh, maybe I can collaborate with these people to do things that I don't know how to do." As a result, we've taken opportunities to work with people who have reached out to us. So there's been word of mouth that we could be people who could collaborate with them and that's provided connections as well-

JW: I want to backtrack just a little bit. You said earlier that once you start doing a change project, then you fall into other change projects. I agree, and I believe you started your career at UT with what I would call a change maker mindset. Why do you think you possess that mindset? Why do you find that interesting? Why aren't you scared by the risks?

RME: I can probably answer this in a couple of different ways. One, what motivated me to get into engineering education to begin with was my dissatisfaction with my own undergraduate education. I thought as I went through many classes, this can't be as good as it gets. This could be better. So I already had this mindset that there has to be a better way to educate engineers. So that motivation to get into the field, I think definitely plays into my mindset now. If that motivated me to get into the field, it also motivates me as I continue in the field. If I thought somebody else could do things better, then I definitely should think that I can do things better and that everybody else can do things better too. So the only way to do things better is not just continue to do it the same way you've done it before. You have to be willing to try new things and see what improvements you can make. So I think that that's one area.

The second area is I have an engineering background and I have a manufacturing engineering background that is wholly rooted in process improvement. As a working professional, the vast majority of my job was improving what was there and how to make it better, how to make cheaper with less defects, safer. That was my daily job. For me, there's a logical transferability of that idea from the manufacturing floor into what we do in education. It just makes sense to have that mindset.

I think that third thing is, I guess, from a personality standpoint, I don't know, maybe I'm weird, but I have a lot of friends who say, "Oh, I don't like conflict. I don't like it when we have disagreements." I look at that situation differently. I think the only way that you can make progress is if you identify where you agree and where you don't agree, and then you try to work through that. This reflects a core idea about change. You should never be happy with the way things are. Things can always be better. So, the idea of change doesn't bother me. I think it's just so natural that it just doesn't bother me.

In fact, to me not changing is boring. It's boring. I don't want to do the same thing over and over again every single day for the rest of my life. That's boring. I want to, at every opportunity, try new things, see if they're better than the things we were doing before. I just don't want to be bored.

JW: Did you ever have a colleague or a mentor say, "Hey, Rachel, you really need to back off of this or you need to pick a different path." I got this advice early on in my career: "Don't talk in a faculty meeting until you get tenure. Don't do these things." And I think that's bad advice.

RME: I agree that that's bad advice. I don't think I've ever had anybody tell me that and maybe that's part of my neuroticism is I've never had anybody tell me that. No one's ever said it, so I'm just going to keep doing it. I know I can be overwhelming to people at times. I can be overwhelming and I'm definitely opinionated. I don't understand when people say they don't have an opinion about something. That just doesn't compute (I have that robot voice, saying "does not compute," in my mind.) I don't think I've ever had anybody tell me to stop working on change projects. Maybe I'm just really fortunate to have had people that I have worked for, or who have mentored me along the way, who have never put that roadblock up in front of me.

The person who suggested I get into engineering education was my boss as a graduate student in my master's program. He was always very encouraging and was willing to listen to crazy ideas that I had, always very encouraging. My engineering boss when I worked at the plant was always very encouraging and motivating and the boss that I have now, who I just took over the position from, he transitioned out of the director role, but is still sticking around to teach. He's always been encouraging and motivating. I think he had to learn to work with my opinionated style, but he never discouraged me. He may help me recraft how I communicate some things, which was good and I learned a lot from that. I learned a lot from listening to him and watching him in meetings and thinking, "Oh yeah, I would've said that in a totally different way, and it would have been bad." So I should listen to how he does that. But I don't think I've ever had anybody say, "Stop talking, don't say that, pull yourself back."

JW: You can be your authentic real self and that's okay too. I like that you gave the example of the former director who would be communicating in a particular way, which made you think, "Oh, I would not have done it that way, but that's a model I can learn from." Were there any particular areas, skills, things that you felt you needed to work on as you were putting your change project together like, "Oh, I'm not sure if I am really able to understand someone's campus culture or departmental politics?" Were there any particular areas? It's okay if there weren't, but if there were anything that you thought, "I need to learn more about this. I need to go look at a website, read a book, something to help me with this," that would be great to hear about.

RME: I am currently reading through a stack of 12 books on leadership as a way to make myself more comfortable with the idea of leadership, but I think some observational things that I definitely needed to watch and learn were one, who are the people on campus who I have a good relationship with in order to ensure that I can get the things done, people I need to partner with but who may not be obvious? For example, the staff in the Registrar's Office is an incredibly important group to be in good connection and in good relationship with. My former boss could send one email to the Registrar if something was screwed up, and it would be fixed because he had a good, long standing relationship with people in that office. He knew how to express his appreciation for them, but he also knew how to clearly communicate what our needs were. He could also suggest some solutions, like "Could we try this? Or could we try this? I've looked for these rooms, would these be options? Or could these department heads be flexible here."

So I could watch him in order to understand who to have the relationship with, and then how to communicate with those people to make a potentially difficult conversation as easy as possible. Being the Registrar may not be a glamorous job and it's a really hard job, but when things get screwed up, they really get screwed up very quickly. So to do things that can be as helpful as possible in that difficult situation can move a conversation forward. I've learned about a lot of different relationships I need to make sure I'm minding in this job for whatever time that I'm in it. My former boss also takes the same approach with our finance and HR offices as well, and I have been working over the past few years to build good relationships there as well. I think relationship building is a huge thing. So watching people that are good at relationship building and sustaining relationships is really important. People can do things for you and help you move things forward if they know you and you know them and you know how to operate with them. That can be a big benefit.

JW: That connects to my next question. If you had a chance to give a piece of advice or a couple pieces of advice to someone who's just starting out, in addition to relationship building that you just talked about, do you have any other ideas or advice that you would give?

RME: I think the other thing that I would suggest if you're brand new, starting out, and starting out especially in a change area, use your beginning time wisely to ask the "ignorance" questions, because when you're new in a role or new in a function, you have a grace period of time to ask questions out of ignorance because you're new. You can say, "I'm new in the job, so could you clue me in on how this works or what's the relationship between these people?" And it's acceptable.

Once you're in a job for a year or so, it seems strange when you start to ask those questions, and people may think that there are ulterior motives when there may actually not be. You may just not know. If you've been there for a while and you ask that, they either will think there's ulterior motives or think you don't know what you're doing. But in that first year, it's okay if you don't know what you're doing, and you can actually learn a lot about culture utilizing those ignorance questions.

Without divulging too many details, I did this the other day. There is a staff member who is mainly associated with our unit. This person also provides a small amount of support for another unit as well, but I wasn't sure about the amount of time this person is utilized. So I took the opportunity to say, "Oh, Ms. X, since I'm new and I'm just trying to figure things out, could you explain: what was the agreement about how this person would be utilized here versus there? How many hours a week are they expected to work in each place? What types of things are they supposed to do for each one? Could you just communicate your understanding of that?"

I actually knew from my previous boss that some tasks and time had been outlined, but I wanted to make sure the expectations matched, and it turns out, they did. It was good. I got a very comparative response and went, okay, that makes me feel good that there's not a mismatch in expectations here about what this person should be doing. The conversation also helped me understand the department culturally, how to work with this other person better since now I knew how they would respond. I didn't want it to be sneaky or anything, but it just was an opportunity that I had to ask that and at the same time, try to build a relationship with this other person by going to them and saying, "Can you line out for me what the agreements were." So I tried to utilize it for a couple of different reasons.

JW: Do you think that people new to a position are reluctant to ask questions?

RME: Yes, I think it's natural for people to have a fear of appearing unprepared. They may think, if I ask a question, especially in the beginning, people are going to think that I don't know what I'm doing. What's important to know is of course, you don't know what you're doing. You're new. You're new, and they don't expect you to fully know what you're doing, especially given how new you are in the role. There's a differing level of understanding of what the expectation was of how much you should know how to do. Yes, they don't expect you to know everything. People don't expect you to know everything and if you don't ask the questions, it actually becomes detrimental later on because if you make it up or you just create something, then people think, "No, that's not right" versus just taking the opportunity to ask so that you get the correct information. It is a great opportunity for relationship building.

When you just say, "Hey, there's information that you have that could be really beneficial to me, and I realize that you have expertise in a particular area. So I'm going to ask for you to share that expertise with me so that I can do my job better" and then we've built a relationship so I know we can work together. That's such great opportunity that you don't want to miss. So you just need to ask the questions. If somebody is mad at you for asking a question, you can be like, "Whatever." Seriously, the vast majority of people won't be, and if they are mad at you for asking the question, that might be a person to stay away from.

JW: You're asking people to talk about themselves and the work that they do?

RME: They love that. They do, especially the faculty members, they love that.

JW: What question were you hoping I was going to ask you about change or doing the work that you're doing, that I haven't asked you yet?

RME: I guess it's always important to think about what happens when you're going down a path of change and maybe it starts going in a direction you weren't expecting. So, you're doing hard work to try to see a process go through and then what happens if it doesn't materialize in the way that you expected it, how do you shift? How do you recover from that? How do you still make it as positive of an experience as possible even if it heads in a direction that you weren't expecting?

JW: So are you getting close to even saying what happens when your change project fails or it breaks, or you thought, dang, this is going to work, it doesn't work. Is that kind of where you're going with this?

RME: Yeah, I think so. I guess part of my hesitancy to refer to that as "failure" is that failure is such a subjective word, failure and success really are so subjective. Failure can have a negative connotation where maybe it shouldn't have a negative connotation all the time. Maybe it's failure and then, okay, pick up and just start over, go to the next step. This failed, so now what's the next step in the process, instead of looking at it as it's over, it blew up, and it's done.

JW: And I'm never going to try to make change, I've learned my lesson. I failed, so I'm done. I'm going to turn in my changemaker spurs and I'm done with it.

RME: Yeah. I need to check on those changemaker spurs.

JW: Hey, wouldn't it be great. Just put those on and walk down the hall letting those spurs announce your arrival. I bet you'd scare a few colleagues that way.

RME: I guess that's what I'm asking. I think an important question to think about is, when you encounter failure, what do you do or when you encounter success, what do you do?

JW: Well, if I follow up on what you said earlier, you might succeed with this project but that spins into another, another, another, another. I think this is how you're going to operate as an academic. This is going to be your life. Welcome to your life, Rachel.

4 Maturing

INTRODUCTION

When you consider the Maturing stage of your change project, you may be tempted to compare it to other aging processes, for wine or cheese perhaps, or reaching middle age. Either of these is appropriate, but perhaps the best image for this phase of your change project is the wave as it forms in the ocean, before it breaks onto the shore (see Figure 4.1). Ocean waves contain two types of energy: kinetic energy, through the motion of the water; and potential energy, from the elevation of the water. Levels of energy in each wave vary depending on a number of factors, and you can see these differences in the power of the wave as it forms. A wave's power can be monumental, a challenge to even the most experienced surfers. Or it can be a gentle splash that tickles a toddler's ankles and causes her to shriek with delight. As the wave matures, it gathers into itself more and more of its power. By the time it reaches you as you stand with your toes in the sand, it has undergone important changes. In all respects,

FIGURE 4.1   Maturing your change project. (Used with permission of the author.)

DOI: 10.1201/9781003349037-4

your change project resembles the wave as it matures and rolls toward the shore. And unlike the Initiating stage of your change project, the Maturing stage requires a new set of tools to add to your toolkit.

In this chapter, you'll explore the Communication, Teamwork, and Diagnosing Problems tools that are unique to this stage of your change project:

- Now that your project is past the startup stage, it's time to get the word out and tell the story of the impacts your project has been able to achieve thus far.
- Now that your project is moving forward, you may also encounter resistance to your project from unexpected quarters.
- In addition, your team will need your attention, not only to celebrate success, but often to remedy interpersonal conflicts that may arise.
- In terms of identifying and diagnosing problems, you'll need to take stock of what is going well, what isn't, and how to address challenges before they get out of hand.

The Maturing stage of your project can be an exciting time, as the wave grows and rises. Just be ready to grab your surfboard and ride the wave!

WHAT'S THE BUZZ?

Tool Type: Communication

Summary: As you tell the story of your change project and its impact on stakeholders, you employ an effective way to communicate about your project at the Maturing stage.

Change makers need to prepare stories about their projects and ready themselves to tell those stories to targeted audiences throughout the life of their projects. This suggestion poses two clear challenges for most change makers:

- **First, what story can you tell about your project?** This question is a matter of focus: it makes no difference where you are in your project, whether you are a nascent change maker or someone who has been engaged in change work for years, there are stories to tell about the work you do, the impact of your projects on individuals, and the potential of your effort to make academic environments better for others.
- **Second, who will you tell the story to?** This question suggests self-doubt: in asking who they will tell their stories to, change makers are making a key assumption. They believe that no one will listen to the story they wish to tell. These change makers may have tried to tell their story and encountered resistance or rejection. It's also possible that they anticipate resistance without ever having tried.

So let's reframe these questions, thinking instead of how we react when a colleague comes to us with a story of their own. Think of a set of specific listeners who want to hear your story. They are eager to know what you are working on because they are struggling with their own challenges in the classroom, in the department, and across the campus. They are as excited to hear your story as you are to tell it. They want to know, "What's the Buzz?"

Here's the buzz:

- The changes you are planning to make, or are working to make, or just made.
- The buzz is what you hope to achieve, or what you seem to be accomplishing, or what you actually achieved through the project.

Why assume that no one wants to listen? If a colleague were embarking on a new project, wouldn't you listen, if only to siphon off some of their excitement to restore your own academic soul? If you open yourself to a positive possibility of telling your story, then what remains is the task of writing that story. Your task as a story teller is to tell a compelling story, and to do that, you have to convey the impact of your project on a person: student, faculty member, staff member, administrator, or even yourself. Your goal is to be brief, descriptive, vivid, and memorable.

PUT THE TOOL TO WORK

To start, let's focus on the end. It is helpful to have the final sentence of your story in draft, because that final sentence will help you focus the story you tell. What do you want your listener to walk away with as a feeling or a conclusion when your story ends?

The final sentence of my story is:

Now that you know where you are going, it's time to do some brainstorming about what the story content might be. To help inspire your thinking and your writing, consider the following prompts:

What were your aspirations for your project when you started out?
The list of aspirations might be long, so for the purpose of this tool, limit yourself to 3.

1.

2.

3.

What moments of insight and revelation have you witnessed in your project thus far?
These insights might be related to your aspirations, so again, limit yourself to 3.

1.

2.

3.

What have you learned along the way?

This is a question about impact. You could focus on the impact you have seen on students, on faculty, or any of your other stakeholders.

From the brainstorming, select one story node—an aspiration, an inspiration, a learning—and begin your draft. Imagine you are telling this story to a good friend over a cup of coffee, and this friend is eager to hear what you have to say.

My Story Node is:

MY STORY DRAFT

Time to start writing! Your story draft may require many words to get your ideas out, but for the purposes of revision, you'll eventually limit your story to 150–200 words. The limit is meant to help you focus on the essential, vivid details. You'll spend less time circling around the story and more time on the story of impact on the individual(s).

Even as you are drafting your story, consider where you are going to tell this story and to whom. Here there are so many possibilities: in a casual hallway conversation, in a committee meeting, at a socially distanced or outdoor gathering of colleagues, in a blog, on a social media site, on the department webpage, in a funding application, to a potential project donor, and the list goes on!

I plan to tell this story to this person/these people within this timeframe:

Why does this person need to hear my story?

```
┌─────────────────────────────────────────────────────────────┐
│                                                             │
│                                                             │
│                                                             │
│                                                             │
│                                                             │
└─────────────────────────────────────────────────────────────┘
```

As you have seen in previous communication-focused sections, your choice of details may vary in order to accommodate the interests and values of the person or persons you are telling the story to. If, for example, you plan to share your story to a group of alumni, you may choose to focus on details that are of concern to them, such as the reputation of the program or preparation of graduates for employment. If, however, you are addressing one of your state representatives, you may need to shift details to reflect the concerns and values of a very different audience.

READ MORE ABOUT IT

I am indebted to Janece Shaffer, founder of Story Ready, for the story approach that inspired this section of the book. Janece works with individuals and groups to write and share their stories, and she encourages participants in her workshops to use storytelling techniques in professional contexts.

MY ACTION PLAN FOR THIS TOOL IS:

TIME BOMB

Tool Type: Communication

Summary: As your project matures, you may encounter resistance from others who see it as a potential threat or challenge. This tool can help identify the sources of resistance and counteract them.

It's inevitable. It may happen when you least expect it. It may come from a quarter from which you expected support, not criticism. And when it happens, it won't feel good. One day your wonderful, innovative, transformational change idea is going to be shot down.

- You may be sitting in a department meeting, confident that the numerous people you talked to about your project will come through in this public setting.
- You may be asked to join a meeting of administrators, including the dean who encouraged you to spend time and energy on this project.

No matter where you are or who is gathered around you, the moment when you hear objections expressed may feel like having the rug pulled out from under you. You may also expect to hear others in room, ones who expressed their enthusiasm in private, to raise their voices to put down these objections, but they remain stubbornly silent. In that moment, you may think that this entire change project idea was misguided. You might return to this book and demand a refund. Why weren't you warned that this might happen?

If you recall, however, there was a preview of this situation earlier in the book, in the Dear Colleague letter. As your project matures and begins to assert influence, you'll need to prepare for the moment when a few solitary voices, or even a small group, raise objections, because it means that your project may be seen as a threat to others. Granted it is difficult to hear your project idea trashed and your hard work discounted. If you aren't prepared for it, the moment can seem like the end of your project. But there is an alternative. It is likely that you will experience this or a similar scenario in a large group setting. How you prepare for this scenario is up to you.

The authors Kotter and Whitehead (2010) have written extensively about the principle of "buy in," a persuasion strategy that helps the change maker diffuse objections. What is clear from them, as well as others who advocate for gaining buy in, is how the principle stands in opposition to creating a shared vision, which we addressed earlier. You may not be a fan of seeking buy in per se, but you may find value in using tactics for disarming the few voices that are willing to speak up to put your project down.

Imagine if you will the large meeting setting, with your colleagues assembled to review your change project. Now take a moment to replay the objections that you heard, objections that were meant to blow up your project, reveal its inherent flaws, and send the project and you packing back to your office. Were any of the objections you heard like these?

- "We tried this in 1992. It didn't work then, and it won't work now."
- "If you try this, you will destroy our department."
- "We can't do this project now. We don't have the money/time/resources."
- "Your change project has merit, but I don't believe you have the expertise to make it successful."

When you hear any of these objections, you may be tempted to dig into your data tables and argue vehemently in defense. Don't.

Consider these objections as versions of a negative comment on student course evaluations. For many of us, student comments that are positive and encouraging, no matter how numerous, pale in comparison to one or two negative comments, such as "Dr. Williams grades too hard," or "Dr. J wasn't in her office when I stopped by." Of course, you can spend quite a lot of energy to defend the fact that you grade with rubrics that are explained and re-explained throughout the term. You can also point out that you respond to email and texts within ½ hour of receiving them. Yes, you can spend your energy defending against the comments, but why would you?

When you review student course evaluation comments, you can distinguish between those that are substantive—ones that you can use to improve the course in meaningful ways—and those that are not. Likewise, you want to identify substantive objections you receive from colleagues, objections that address important concerns that you should consider and possibly address to ensure success for your project. The objections that you may hear from colleagues, like those listed above, are the professorial equivalent of one-off negative student course evaluation comments. Don't expend much time to address them.

The public gathering setting does, however, offer you an opportunity to address others who are in the room, not the one or two commenters who wish to shoot you down, nor the vocal supporters who you can count on to remain in your corner. Instead, your focus should be on the other colleagues—the "persuadables," for example, the ones who you believe can be persuaded—who won't speak up during the meeting and who may lean toward supporting your change project but who wish to remain silent to see how you react to being challenged. In this regard, your best bet is to resist the temptation to spend your time trying to convince the naysayers and instead turn your attention to those who could be persuaded:

- A colleague points out that the department tried something similar in 1992: you can point out to the "persuadables" that your college and students have changed a lot since that time, and so there is an opportunity to make a positive change for the new generation.
- Your expertise is challenged, someone points out that the conditions for change are less than ideal, and so on. You can be ready to respond to these and other baseless objections.

The important thing to remember is not to take the bait. While those who object to your project may be testing you by challenging your ideas, they may also be tossing you a time bomb set to blow up your project.

PUT THE TOOL TO WORK

You can approach this tool from two different directions. You can start with the objections that were raised, or you can engage in a bit of speculative writing and imagine the objections you could possibly receive. Either way, take a moment to catalog the objections and determine whether they are substantive. For example, one objection that cites lack of resources may be substantive, while another is not, such as the objection that comes from the colleague who reliably talks about resources without having a good grasp on the current state of financial affairs on campus. When you have distinguished between the substantive and the non-substantive, draft out a possible response that is directed at the "persuadables."

Objection	Substantive/Not Substantive	Possible Response

Take the time to not only write out your response but practice the response so it comes readily to mind in the emotionally charged setting of the meeting. Having your response ready will help you stay calm and less likely to rise to the bait. And less likely to be shot down.

READ MORE ABOUT IT

Kotter, J.P. and L.A. Whitehead. 2010. *Buy In: Saving Your Good Idea from Getting Shot Down*. Cambridge, MA: Harvard Business Review Press.

MY ACTION PLAN FOR THIS TOOL IS:

TENDING TO YOUR TEAM

Tool Type: Teamwork

Summary: A change project's success depends on the success of each team member. By tending to your team, you can ensure that each team member is working at a high level while feeling supported and valued.

> Nobody heard him, the dead man,/But still he lay moaning:
> I was much further out than you thought/And not waving but drowning.
> Stevie Smith, Collected Poems and Drawings (1983)

Everyone has a colleague who seems to need no one's help. Theirs is the curse of competence. There is nothing this person cannot do. Collaborate with them on an article, and before you have even put pen to paper, they have written the first draft. Agree to co-present with them at a conference, and you can be sure that the PowerPoint slides are finished weeks before the due date, while you are still figuring out which hotel to book. Working with this colleague is an unmatched joy. You never worry that your colleague will neglect assigned work or an upcoming deadline. You don't need to be concerned that your colleague will ignore your urgent email. For this, and for all of the other strengths that your colleague exhibits, their proper superhero name should be The Dynamo.

Their alternative superhero name could also be The Intimidator, since you may find yourself from time to time intimidated by the sheer level of accomplishments and quality of work that The Dynamo is able to produce. It appears perhaps that The Dynamo doesn't need or want your help. You could just step back and give The Dynamo room to write the article and create the presentation and the superior quality of their work would reflect well on both of you. Perhaps that is what you Dynamos prefer. But even the Dynamos among us may need help from time to time. Like the dead man in Smith's poem, what you think is waving may be drowning, and you need to keep your eye on that swimmer who is far out. The Dynamo may be a strong swimmer who has made the choice to dare the rough water, but there may also be the chance that the Dynamo is in need of a rescue.

Tending to your team requires that you can recognize the specific superhero alter ego that best characterizes this individual and their unusual gifts. And the best place to start to understand the gifts of others is to assess your own. Let's start in the world of female superheroes and some of its most iconic women. Wonder Woman rescues humankind from the ravages of Nazi ideology, and her hourglass figure and golden lasso are important assets. Storm's superpower is the ability to freeze her enemies. Maybe that's not a power we would envy, but you can see its uses in particularly sticky situations. Department meeting not following the stated agenda? Irascible colleague taking too much time rehearsing the same rant you have heard time after time? Freeze them!

My own superpowers are bit less worthy of a comic book but still I am proud that I can unleash them on unsuspecting opponents in order to serve worthy friends.

During a department retreat, I asked each staff member to create their own alter ego in the form of a superhero, hoping to use the exercise as a way to accomplish a few objectives. I wanted each person in the meeting to identify a few of the skills and strengths that they believe they bring to their role. This identification was not just for their own benefit, but for the education of other staff who needed to be reminded of the talents of their colleagues and encouraged to value those talents. I also hoped to give each person a chance to reflect on their talents and understand that these talents were ones that could be used for the benefit of others. The result of the exercise was remarkable. The staff member who felt that at times his skills in data collection and analysis were overlooked created his alter ego Data Man, complete with cape and a fistful of data tables clutched in his right hand. Another person highlighted their skills in empathizing with the feelings of others with their alter ego representation, all focused on emotions. For me, the alter ego turned out to The Bridginator. My superpower is the ability to connect people across disciplines in order to encourage new, productive collaborations. And I have a story to tell about my life as The Bridginator.

My superpower comes in handy at that familiar and often dreaded event at face-to-face conferences: the networking social. I think many people find these occasions a source of anxiety: a hotel ballroom crowded with people you don't know, with the expectation that each of us is supposed to join random conversations, make witty comments, and balance a small plate of mushroom caps, mini-quiches, and boiled shrimp in one hand while impressing total strangers. Such social situations are difficult for many of us, but at one particular conference, I donned my disguise as The Bridginator and found a new purpose. Also attending the conference for the first time was a colleague from a different institution, a person I had been working with on a project unrelated to the current conference and someone who knew far fewer people than I did. Rather than focus on the people I didn't know, I focused on introducing my colleague to the people I did know. I knew that he needed to make connections at this conference. There were many people in the room who could be potential collaborators for him, and we spent some time helping him get to know these new connections. At one point, he seemed to recognize what I was doing in my role as a bridge maker for him, and he commented on it. Rather than consider my anonymity compromised (The Bridginator doesn't wear a mask, just so you know), I saw this as confirmation that I was doing good service in this community, and I was encouraged to consider this my special skill, my superpower.

PUT THE TOOL TO WORK

In order to craft your superhero alter ego, take a few minutes to reflect on your special skills and unique talents that could comprise your superpowers. Are you the person at department meetings who can skillfully summarize the rambling arguments of others? Do you have a knack for scrounging through budgets and finding untold treasures? Have you been told that you are great at a particular task like creating bold and beautiful spreadsheets? No matter what the skill, please consider this your superpower.

My superpower(s) are:

From the starting point of your superpowers, create a visual representation of your superhero alter ego. For those with artistic skills, this is a great time to break out your supplies and really have at it. For those with more limited abilities, you can take a superhero figure and adapt it for your own uses. The Bridginator hangs near my desk as a figure on a sheet of white paper who is clearly Wonder Woman but with a purple and green costume that reflects my own color choices.

With a superpower defined, it's time to consider when and where your superhero should be sharing their gifts in the context of your change project.

- What is the next situation or context where your superhero could make an appearance?
- What difficulties or challenges could your superhero remedy?
- What would be the reaction of your colleagues to the arrival of your superhero?

Creating a superhero alter ego is also a task you can do with individuals who are working with you on the team. They are also in possession of remarkable gifts, talents that you are asking them to deploy in order to move your change project forward. By encouraging them to identify their super powers, you have made their unique contributions visible to other members of the team. And when a team member struggles, like the man in Stevie Smith's poem who isn't waving, but drowning, then you have a strategy with which you can identify the struggle and offer a lifeline.

Read More About It

You might find it easier to consider superheroes from the perspective of those less than heroic. If that is the case, then consider watching *Mystery Men* (1999), with Paul Ruebens gifted with an awesome, albeit noxious, super power.

My action plan for this tool is:

KEY PARTNERS AND COLLABORATORS

TOOL TYPE: TEAMWORK

Summary: At the Maturing stage, you may find it necessary to reach out to find partners and collaborators who can help you advance your project beyond your local sphere. The Key Partners Map tool can assist in identifying potential partners and collaborators and measure their support and influence on your project.

Change makers are as diverse as the change projects they undertake. Some are content with implementing change in their own classroom with the courses over which they have almost complete control. Others are happy once they can see the results of their change efforts impacting colleagues in their home departments. While these descriptions may reflect your current goals as a change maker, you should also consider what you could achieve with your change project if you reached out to new collaborators and partners.

Rather than searching for collaborators and partners without a clear process, you can benefit by employing the Key Partners Map tool outlined here. It will be of particular use to you if:

- You have been successful with a proof of concept of the change, and you think that other departments on your campus would be interested in adapting it to their own uses.
- Your project has the potential to impact stakeholders outside of your department or college and could benefit from a partnership or collaboration with non-academic offices, such as career services and student affairs.
- You envision your change making an impact with other departments on other campuses across your state, your region, your country, and/or the world.

In order to extend the reach and influence of your change project, you will need to consider potential partners and bring them into the sphere of your project. There's just one problem: how do you identify the right partners for your project, how do you approach them, and how do you know they will sign on? Creating a Key Partners Map will help you determine your next moves.

Before we jump into the tool, however, let's take stock of your project as it is currently operating at the Maturing stage:

1. What challenge(s) first caught your attention and inspired you to work on this project? Take a moment to write down a few of these challenges.

Consider a few examples. You may have been struck by a particular gap in students' preparation for more advanced classes, or by the lack of targeted mentoring for faculty in your department, or by the difficulties alumni of your department had with connecting with their college after graduation. These are only a few examples to illustrate, and the challenges that inspired you to develop your change project will be specific to you and your context. It is important to recollect what motivated you to remedy the challenge, since the same challenge may be a concern for the potential partners and collaborators we are seeking.

2. Now that your project is in the Maturing stage, what have you learned about the challenge(s) and solution(s) you implemented? Take a moment to write down a few of the lessons learned.

The lessons learned are key insights that your potential partners will also need. They will naturally have questions about why you chose this particular solution to the problem, and you can be even more persuasive if you can highlight both the accomplishments and travails of the change work you have done thus far.

Okay, now you have your thoughts in order. Let's go going with the tool. As you use the tool, you'll be identifying potential partners and collaborators who can join you in your change project. Let's use the following example to highlight how the tool can help you recognize where the opportunities lie.

PUT THE TOOL TO WORK

When you create a Key Partners Map for your change project, you are using the tool to both identify a potential partner for your project and to visualize the connections between your project and the needs or interests of the partner. Begin your analysis by creating an inventory of partners you have already established a collaboration with. You can use a broad definition of "collaboration" here, everything from a partner who is working with you to implement a component of your project, to a partner who is at the initial stages of such a working relationship.

Partner—on- or off-campus	Department/Area/Office	Their role or impact on your project

Consider not just the partners you are already working with, but take a moment to reflect on partners you have thought about working with but have not yet approached. For example, you may have considered approaching representatives of a local industry who could serve as mentors for your students, or you might have toyed with the idea of connecting with a national organization that is focused on a similar need that your project addresses.

Partner—on- or off-campus	Department/Area/Office	Potential impact on your project

As you analyze potential partners, you should be asking yourself, why? Why should you approach this partner in particular? What impact do you envision resulting from a collaboration with this partner?

Now select one or two partners with whom you would like to initiate a collaboration and think more deeply about your motivation for partnering with them.

What motivates you to pursue a strategic relationship with this partner?

1.

2.

Strategic partnerships are key to moving your project forward and ensuring its sustainability. The challenge for many change makers is twofold:

- How do you identify the right potential partner(s) for your project?
- How do you communicate with that partner(s)?

An example of these two principles may help. As part of the National Science Foundation RED program, RED teams were asked to consider their projects in the context of identifying potential partners. This task inspired several teams to visually depict their RED project in terms of a map using a metaphor that tied directly to the discipline of their project. For instance, the environmental engineers depicted their current and potential partners as part of a system of streams and rivers all feeding into a project that served undergraduate students successfully. Yet another RED team represented their core team at the center of their map with each successive circle indicating another partner who surrounded the core. There was even a map that showed potential partners as circuits in a bus, and this was definitely not a transportation metaphor!

As you begin to create a Key Partner Map that is appropriate to your project goals and your institutional context, you are beginning to identify not only the potential partners you wish to work with, but also the individuals currently working on your project who have connections to these partners. As a feature of your Key Partners Map, identify a connection between individuals on your project team and the individuals/offices that you see as potential partners. For example, one of the faculty working on your project team may have an existing partnership with the Admissions Office or the Registrar that predates your project. It is important to remember that partnerships begin with relationships between individuals, not groups. Effective change agents take advantage of opportunities to invest in relationships. From the institutional perspective, resources including technology, information access, expertise, control over decision-making, and space comprise capital that can be contributed to or requested from partnerships. Bringing the interests of the partners into alignment, along with the capital they can contribute, generates forward progress in change efforts.

Every change project at the Maturing stage consists of data/tools/insights/personnel, etc., that represent something of possible value to a potential partner. Consider your project's items of value from the perspective of potential partners. At this point in creating your Key Partners Map, create an inventory of these items. You are speculating on what your potential partners might value. I have included one example of a product of value and how a potential partner may envision the benefit of using that product.

Item of Value (specify type)	Who Might Want It? (position or office)	What's Their Interest? (main demands facing this position/office)
For example, your project may have created a climate survey that was used in your department.	Another department on campus may have a use for a survey like yours.	The other department may need a survey like yours because they have some of the same issues with student recruitment and retention.

Likewise, every change project needs resources that can't be supplied solely by the team itself or the department. Consider your resource needs from the perspective of potential partners. You may find it helpful to consider potential partners both on and off campus.

On-Campus Resource (specify need)	Who Can Provide It? (position or office)	What's Their Interest? (main demands facing this position/office)
For example, your project may need admin support.	The dean's office could provide it.	The dean's office' could see a partnership as beneficial for them since it could improve personnel management, or assist in the administration of the academic mission of institution.
Off-Campus Resource (specify need) You may need access to individuals with industrial experience.	Who Can Provide It? (potential partner) A local industrial partner could offer individuals with this experience.	What's Their Interest? (main demands facing this organization) A partnership with your project could help that industrial partner find students to recruit, ensure quality control, etc.

As you begin to include details on your Key Partners Map, you should prepare for the communication work that will be involved when you approach potential key partners. Communication plays a key role in approaching a potential partner and establishing a strategic relationship. Given what you believe you can offer of value to the potential partner, draft a brief talk (five minutes) in which you propose a collaboration with the partner for your mutual benefit. You can envision this talk as part of an initial conversation you have with the partner, or you can develop a talk that would occur further into the collaboration. Use the following prompts as a guide:

Define the challenge as you see it from your perspective:

Offer your own understanding of the partner's needs, acknowledging that your understanding represents an outsider's perspective:

Create a picture of the future that is improved as a result of your partnership:

How would this talk change if you gave it to a different potential partner?

As is the case with other communication tasks associated with your change work, practicing the talk will be beneficial. You can try it out with friends and colleagues who are supportive and predisposed to like the idea to begin with, or even seek out colleagues who will challenge your perspective. When you think you are ready to take the talk on the road, be sure to identify the right individuals to approach. In some cases, you may be familiar with specific individuals, or you may have to theorize a potential partner based on their role or function, rather than knowing this individual personally. You may also find it useful to organize your targets by the office, department, or division where your partners might work. For example, you may know that the associate dean for student affairs is Alice Smith, so you can record her name as well as her office. Ms. Smith could be instrumental in helping you connect with other departments who have been seeking new ways to support incoming students and improve retention. For other potential partners, however, you may know vaguely that your college has an office dedicated to coordinating with community service organizations, but you don't have any idea who works in that office. In either case, you can research to make your approach to a partner as concrete as possible.

Even as you build your Key Partners Map and solidify partnerships for your current change project, you should strive to explore a wide variety of possible partners and collaborators. Since each change project needs its own specific Map, you may find yourself identifying different individuals and their roles as you map different change projects. And when you are ready to begin your next change project, your map will help to guide you on your way. If you invest time to plan potential partnerships, you will understand what you need to do next, what questions still remain, and who to approach first and when. And don't forget the additional value you gain from making your Key Partners Map: it's a vivid graphic you can hang on your office wall, serving as a daily reminder that your change project needs partners who can offer their support and influence while benefitting everyone.

READ MORE ABOUT IT

In addition to the Key Partners Map tool, I recommend taking a look at the following two resources. Froyd et al. (2015) and Eddy and Amey (2014) present useful approaches for identifying and connecting with potential partners, and they have the research data to back up their claims.

Froyd, J.E., Henderson, C., Cole, R.S., Friedrichsen, R., Khatri, R. and Stanford, C. 2015). From Dissemination to Propagation: A New Paradigm for Education Developers, *Change: The Magazine of Higher Learning*, 49:4, 35–42, DOI: 10.1080/00091383.2017.1357098

Eddy, P. and Amey, M. 2014. *Creating Strategic Partnerships: A Guide for Educational Institutions and Their Partners*. Sterline, VA: Stylus Publishing.

MY ACTION PLAN FOR THIS TOOL IS:

CATCH THIS VIEW

TOOL TYPE: DIAGNOSING PROBLEMS

Summary: Borrowing an artist's tool, you'll take a closer look at the difficulties you may have with your project at the Maturing stage.

If you wish to pursue a long-held dream of being an artist, you might start your career with online art instruction modules designed for kids but are perfectly appropriate for someone who knows very little. For example, the Hugh Lane Gallery in Dublin, Ireland's National Museum of Art, offers art activities via YouTube videos, focusing on different hands-on projects that demonstrate artistic principles. Take, for example, the Summer Camp Day 3 Activity: Make a micro-notebook. The activity is to create a small drawing notebook, but more interesting is **what** we should capture in this notebook. The size of the drawing surface is small, so this is not the place to render a mountain range or a skyscraper.

Rather, this is the right place to depict selective views of things, like Liliane Tomasko does in her painting of an unmade bed. As the YouTube instructor suggests, Tomasko is showing not the entirety of the bed, but a small piece, covers folded back, just that curve and a few straight lines of different widths, a focused view of an everyday object that is transformed into an abstract masterpiece by virtue of preserving one section of the folded spread.

Use of a viewing frame is part of the activity. If you have never used one of these artist's tools before, Blick's Art Supplies will send you your very own ViewCatcher™, "The Artist Tool That Teaches," with a promise that it will help you "become a better artist," along with several testimonials. Skip Whitcomb states "Rarely does an artists' tool come along exhibiting this degree of thoughtfulness to basic problem solving." And the problems it will solve? Composition. Color value. Color decisions. And if it's that good for artists, surely it could offer change makers some help as well, as a way to identify, assess, and remedy problems in their projects.

One recurring difficulty is holding both the project as a whole and the project constituted by specific elements at the same time. You may be asked, what is your project trying to achieve, what are your measures for success? and that question may prompt an attempt to capture the project writ large, big picture. If you consider your change project in its entirety, however, you may miss the project details that represent its degree of impact. What impact does your project have, for example, on a specific individual? Perhaps your project includes a professional development component for students who are members of under-represented groups and who may not have professional role models in their communities. Pick up your ViewCatcher and turn your attention away from the whole project to concentrate on one component, professional development as it is experienced by a student cohort, or a specific student. Now you can see the detail of the program as it is exemplified in Tomasko's painting. Further, you may catch of glimpse of what isn't working for this student cohort or that particular student, which may be an indicator of issues that need to be addressed. Oh yes, the ViewCatcher has a place in your change maker toolkit.

PUT THE TOOL TO WORK

Putting this tool to work doesn't require that you invest $10 in acquiring your own ViewCatcher. Consider your tool as metaphorical, rather than actual. That doesn't mean that it isn't effective as a mental tool.

First, identify one particular aspect of your project that has not been working as well as you originally expected, or a problem that arose in the midst of other successes with the project. Use the ViewCatcher to isolate this specific part of the project and assess current state using some or all of the "views" listed:

People

- Who is engaged most in this aspect of the project?

- Who is impacted directly by this aspect of the project?

- Take a moment to reflect on what the impact has been on an individual or group.

- Are you unsure about impact?

- Can you talk with the individual or members of the group to gain clarity about impact, how they experience your change project in a direct way?

- Have you collected data along the way that could be useful to gain insight on the impact of your project on people?

Process

- How does this aspect of the project operate?

- What processes are currently followed to make this aspect of the project happen?

- Take a moment to reflect on the possible difficulties you can see with the current process.

- Can you isolate specific aspects to correct or remedy?

Products

- What specific outcomes or results have emerged from this component of the project?

- Are these the products that you expected, or are you seeing products that were unexpected?

- Take a moment to reflect on the state of the outcomes thus far.

- If the results are not what you expected, are they still satisfying the goals of your project?

- Do you need to recalibrate goals or revise outcomes in order to restore proper alignment?

READ MORE ABOUT IT

You can find directions for making your own ViewCatcher, as well as other art projects at the "Make a Micro Notebook" site produced by the Hugh Lane Gallery.

MY ACTION PLAN FOR THIS TOOL IS:

KINTSUGI, OR PUTTING THE BROKEN PIECES TOGETHER

TOOL TYPE: DIAGNOSING PROBLEMS

Summary: In the Maturing stage of your change project, there may be problems that arise that can be fixed, as long as you have the right tool to help identify them.

On her website, ceramics artist Linda Fox writes that she draws her inspiration from "the decorative arts and architecture of the Arts and Crafts Movement of the early 20th century," describing her vases, trays, and other objects as the "modern interpretation of the pottery of this era." In order to achieve her interpretation, she uses matte glazes, rather than shiny ones, glazes that keep the viewer's focus on the "strong and elegant form" of the objects. Both the form and the glaze drew me to her work during a visit to the Pewabic Pottery in Detroit. I purchased one of Fox's platters, watched as the clerk wrapped it up carefully, and got it home without incident. My memory of the purchase is vivid.

What is less clear is how I broke it. Did I drop it onto a table, or did it slip from my grasp as I was returning it to a shelf? The platter was in three pieces, not one, but I couldn't just toss it into a bin. I collected the piece into a cloth shopping bag and stuck them into a closet.

On my next trip to Detroit, I returned to Pewabic, hoping to find a replacement platter. The helpful clerk suggested another alternative: kintsugi, the Japanese principle of embracing imperfection through repairing broken pottery. With the necessary supplies—quick-drying epoxy, gold leaf, small paintbrush—and reviewing articles and videos about kintsugi, I set to work. But even as I began the process, I had doubts and questions. First, although I could envision what my repaired platter would look like when I was done, I was afraid I'd do it wrong and the whole thing would look worse, with gold smeared over the entire thing or pieces that won't fit back together as they should. Second, I couldn't help but see relevance to academic change as I mixed the paint pigment with the EC 600 glue (but I had to do this fast because it dries in an instant, and I didn't want to glue my fingers to a pottery piece inadvertently).

As change makers, we may envision our projects in an ideal state. We may insist on unblemished perfection. It is more likely, however, that the change we make won't live up to that ideal. In fact, our projects will not be perfect. They will sustain cracks and need repairs. They will exist in a state of constant forming and reforming, all through their lifetimes.

It's good to remember that, like my damaged platter, the broken project shouldn't be discarded. In the tradition of kintsugi, we have the opportunity to identify the cracks, chips, and breaks, and repair them as they arise along the way during the life of the project. Sure, you can choose to ignore these problems, but perhaps you could take a moment to assess what has not gone well with your project, especially the interpersonal concerns, rather than the errors in design and execution. The former mistakes can create long-lasting damage that no amount of glue can repair. If left unaddressed, these mistakes will impact future projects.

And I bet you are curious how my own attempt at kintsugi turned out. The platter is now in one piece, and yes, the cracks are visible and beautiful. And I am okay with that.

PUT THE TOOL TO WORK

In order to understand the problems that need to be addressed, and to prioritize among them, let's first make an inventory of the cracks, chips, and breaks you can currently identify in your project. We'll use a color coding system, from yellow through orange to red, to indicate levels of severity. A yellow item is a problem, but not a project-threatening one. An orange item puts the project in possible jeopardy. A red item is the most serious and could spell the end for the project and for your future work as a change maker.

Yellow	Orange	Red	What's My Repair?

For each crack, chip, and break, consider what repairs you can effect. For example, if one of the serious breaks you can identify is interpersonal, then propose a few options that could help repair it. You may have a dissatisfied member of the project team who has decided to ignore deadlines and emails. You don't know why this has occurred, and you may have chosen to turn away from this situation in order to direct your attention to other pressing issues. Your repair could be to reach out and ask for a private chat to see if you can uncover the issues that have impacted this person. Their detachment may have nothing to do with your project; they could be struggling under the weight of professional demands, family concerns, or other issues that you were not aware of. A kintsugi repair may not remedy all of the cracks, chips, and breaks you list, but you also have a clearer sense of what is possible to fix through the inventory.

READ MORE ABOUT IT

You can learn more about fixing broken pottery at the Kintsugi Pottery: The Art of Repairing with Gold site: https://www.invaluable.com/blog/kintsugi/. You might also enjoy visiting the site of the artist who made the tray that I repaired at Linda Fox Pottery: http://lindafoxpottery.blogspot.com/. For insight into the history of the Pewabic Pottery operation in Detroit, Michigan, check out their site: https://www.pewabic.org/.

MY ACTION PLAN FOR THIS TOOL IS:

CHANGE MAKER INTERVIEW: MR. THOR MISKO

At the Maturing stage, you may uncover new challenges around resistance to your project, even though the project has been in operation and is making an impact (and perhaps because it is making an impact). Mr. Thor Misko is currently vice president of project development and minority owner of Selzer-Ornst Construction Company and previously served as team leader and program director at the Kern Family Foundation. During our interview, he said something that strikes me a supremely useful for change makers who are contending with resistance: "Why spend your time on that 10% of people who are going to just wear you out, when you have these other people who really want to do this well and can connect and engage?" That is the kind of profound message that many of us need to hear right now.

Thor Misko is vice president of project development and minority owner of Selzer-Ornst Construction Company—a construction industry pioneer building flourishing communities since 1928. Thor had an untraditional path and came to Selzer-Ornst from the Kern Family Foundation where he shepherded their entrepreneurial engineering program efforts to transform engineering at colleges and universities throughout the country. Prior to the Kern Family Foundation, Thor was an executive with Project Lead The Way (PLTW), the nation's largest K-12 STEM education program, a director at Milwaukee School of Engineering (MSOE), and worked in various entrepreneurial and engineering fields and markets throughout the United States. As a building systems engineer, he had the opportunity to work on a wide variety of projects in the education, commercial, healthcare, hospitality, industrial, entertainment, and gaming sectors comprising over $1.5 billion in total constructed value.

JW: What were your primary responsibilities when you worked at the Kern Family Foundation?

TM: Originally I was approached about this opportunity at the Kern Family Foundation (KFF) when I was working at Project Lead The Way. KFF was one of our primary funders, and I had gotten to know a lot of the folks at the Foundation over the years. In 2015, I sat down for coffee with Jim Rahn, and I was already involved peripherally with some of the work that happened with KFF in the past, particularly with the Kern Entrepreneurial Engineering Network (KEEN). That coffee meeting went a little differently than I had anticipated. Instead of conversations about the next steps for Project Lead The Way, Jim talked about making great strides on work KFF had been doing with the entrepreneurial engineering team. He said, we love the work that you've been doing at Project Lead The Way and the relationships that you've built over the years with different higher ed partners and your passion for higher ed engineering. He asked, would you be interested in coming to help lead this team that is driving this work forward and really thinking about what's the next for engineering education across this country? Jim said, we have some ideas and we have a lot of people who are interested in the work. We've connected with them around the concept of "mindset" and "skill set" and how those two components are so critically important for the next generation of engineers to be successful. Would you be willing to help lead that team? If you think about the trajectory of where the current entrepreneurial engineering network was at that time, there's that famous "S" curve, and we were in the upward slope of the "S" curve. We were starting to move in that upward trajectory. We had a lot of proof of concept projects, if you will. We had different things that were happening across the country at different sites that basically were showcasing that entrepreneurially minded engineering was important and having an impact. The dominoes were falling and the larger dominoes were getting easier to knock over because of the fact that early adopters were paving the path and serving as the pioneers.

When I said yes to that opportunity, one of the first things I did was ask, how do we organize our thoughts, ideas and people and essentially our resources to really live into this mission of graduating engineers with an entrepreneurial mindset so that they can create personal, economic and societal value through a lifetime of meaningful work? That mission is something that I think is, obviously, critically important and I continue to be a huge advocate for. That was my directive, if you will.

It was important to figure out how to make an impact in engineering education at the undergraduate level. We saw that we could accomplish the mission of fundamentally changing the way that engineering is going to be taught across the country, by and large. Our plan wasn't to just engage a few schools that are really doing this and doing it well, but rather, there was this desire for every institution that we were talking to really lean into this heavily. It was important to have exemplars who could to help guide the way. What is the proven journey that we can take? Yeah, we're going to make it our own along that way, but we didn't have the heavy equipment to forge the path. We just wanted to make sure we had our own trails, so what were those paved paths that we could showcase and help people understand that no matter what type, size, shape of institution you were, that you could really do this and do this well, and why this was important, why this was so critical, not only for the future of higher ed in my mind, but also, ultimately, for the future of engineering in our country and the way that engineers are going to have an impact on society.

Once those things all started to happen, basically, that allowed me to turn around and say, "I think we can make this work." My role was to help, essentially, organize resources to make the largest impact possible around the future of engineering education in the country.

JW: It does seem to me that this role you were in gave you a view about change in STEM education, specifically, higher education more generally. You enjoyed many different views on many different projects. What do you think was most surprising about what you saw happening in higher education during that time, from 2017 to 2020, or even before that, given your unique perspective?

TM: Most surprising, do you mean in general, around STEM education, or around the systems or the institutions themselves?

JW: I will let you take that. I'm interested in people and institutions and how they might surprise you.

TM: One thing I realized early on when I started working with higher ed was the premise of autonomy. This desire for autonomy within higher ed that was paramount. No faculty member wanted to be told what to teach. They wanted to be the masters of their own domain. Yes, they wanted to understand what the research would suggest. They wanted to understand what other people were necessarily doing, but they wanted to come to their own conclusions. No one wanted their superiors within that system to be basically dictating what something should look like.

That is obviously very, very different than the K-12 environment. Yes, teachers are working alongside administrators, and they're selecting content and curriculum that they want to teach, but there are standards and other things that they're really focused in on that guide what they want to do. So what I started to realize is that, because of that desire for autonomy in higher ed, there was kind of a communication issue that was ultimately happening in higher ed, and not necessarily for the wrong reasons. It was all for good. What I found was that there was redundancy in the content being taught, not on purpose, but accidentally.

In higher education, you're teaching this course and the person who taught the course that was a prerequisite for that course, originally back whenever it was designed, faculty were supposed to be teaching certain things, and over time, you as the faculty member in charge of the next course didn't feel comfortable or confident that those things were being taught to your liking, so you started adding stuff into your course, and as a result, your syllabus changed. It was different than it was when you originally designed that course, perhaps, because of the fact that you wanted to make sure that students were learning those fundamental components so that they could continue to build off of them. But you didn't repeat the content because you were trying to reinforce certain concepts. You repeated it because you didn't feel that the students had a grasp of that topic, that concept, whatever it was.

There was a lot of repetitive work that was happening in higher education because of some of these communication issues, and also because, maybe this course right here is a core course that everyone is taking in engineering or in this specific field of engineering, and so there's now four professors that are teaching these different sections of that course, and so now, depending on what groups of students show up over here, some of them have a better understanding and some have a worse, one and it's not necessarily based on the student. It was also based upon the faculty member that they had earlier. Even though through ABET, you're supposed to line these things up and showcase the topics that are being covered, you don't necessarily usually map them out and showcase how you're reinforcing, how you're building, how you're scaffolding, how you're spiraling out that content and curriculum.

What ended up happening because of that, there was also this thought in higher ed that there was just not enough time to do anything more than what you were already doing, because there was so much that was happening. But a lot of that was not accurate; it was because we weren't having those conversations with one another to say, "Hey, what if we did this differently? These are the things that are so important for our students. These are the concepts that we want to make sure that they understand and are mastering throughout their program. What are we doing to really build that and support one another while we work to do it?"

Those were things that were really interesting to me when I started to understand how higher ed works. I also realized early that there was a lot of misunderstanding around some of the accreditation components within higher ed. They were often used as excuses by administrators and faculty alike. An easy answer was, hey, we should improve the way we teach or do something different in our curriculum, but the first response was often, "Well, ABET." It was like something that just came out of faculty and administrators' mouths, like, "I don't want to do anything over here. ABET." These things would just flow out of people's mouths as ways to shut down a conversation, but they weren't justified responses.

They were just excuses, so it gets to that idea of, what are real and imaginary boundaries that truly exist in higher ed, and I think there were a lot of imaginary boundaries. There were these things that people thought that they couldn't do or they shouldn't do because they didn't want to jeopardize accreditation. They didn't know if they had authority to necessarily make a particular change, so even though there was that desire for autonomy, there's also uncertainty, like, "Do I actually have authority to do some of these things? I don't know, like, maybe." "This is what I was told I'm supposed to be teaching, and I was given a syllabus," which is the other beautiful thing that often happens with new faculty members. They are told, "Hey, here you go, here's this course," and maybe if they're really lucky the person who taught it before is still there so they can talk to them, but if the previous professor isn't there, the new faculty member is kind of on their own. They have some of these materials, and so what do they do? You're stretched thin as a junior faculty member. You're trying to figure out the system and what you're supposed to be doing, and so you fall back on what you know.

They say to themselves, "I took thermodynamics, so, let me pull up my notes and let me see the syllabus, let's put this together and I'm going to teach it the way I was taught," and "I wasn't necessarily an expert in thermodynamics. I got my PhD in "fill in the blank," and, yeah, I took that course in my undergraduate program, but it's not like that was my forte, perhaps, and now I'm teaching that to a cohorts of students that are coming through." I found this intriguing because it's the opposite of what I was talking about before, which was faculty desire for autonomy. You wanted this to be yours. You wanted to have your fingerprints all over the course, if you will, but you also were thrown into the fire and so you were like, "What is everyone else doing?"

In that moment, you wanted to know, and you were scrapping around trying to figure out what everyone else was doing to make sure that course could be even better than what you were given, and so there seemed to be this desire, this need to find resources quickly, even though no one would say it out loud because they always wanted to have their own fingerprints on the course. You wanted to be able to adapt versus adopt. That was a paramount moment for me in higher ed. In K-12, you want to adopt a curriculum. In higher ed, you want to adapt a core basis of something to fit within your culture, within your ethos, within your campus, within your course, and so that, I thought, was just paramount.

At the end of the day, every faculty member is a human, and we all have that desire to do things well, and we want to know and have all the access to the different really cool things that other people are doing so that you could make determinations and decisions around how that's going to fit and work in your coursework. But now you make all those changes and there's no feedback loop, and so now you've gotten back into that communication issue which is what I was talking about in the beginning. So you're like, "Hey, let's adapt this, adopt that," and then all of a sudden, boom, now you're back in that same area where your course is different and likely better than it was before.

The courses that come before and after you haven't necessarily changed because there's no coordination happening in some of those different areas, and now you even have more of that problem of, what's being taught over here was already being taught over here, and by the time that that professor realizes it is usually a couple years when they're saying, "Why are the students coming in and they already understand these concepts? What's going on? Maybe I need to change this." That seems to be where, I would say, the continuous improvement typically happens within courses, usually on the far ends.

It's the things that were going really well, and you want to make some different tweaks on it. You get excited about it because they're the things that are fun to teach, and then the other part which often gets avoided a little bit longer than they should are the ones that are the hard concepts to teach well and help people get their arms around. After the end of that course, you always sit back and you think, "I have to fix that. There's something not right there, and I have to figure out how to do that better because it didn't go the way I wanted it to." Then time sneaks up on you and smacks you in the face and then you're jumping into the next thing and you haven't updated whatever that course work was.

That was another realization: just like other professions, you're busy. You're jumping from one thing to the next, and you have all these different things going on. You have different areas going on that are pulling you, between your scholarship, your service, and your teaching, so you don't have a ton of time to always connect those things together and also to really lean in the way you want in all three aspects. Those were all, I think, some epiphanies I had early on, and then just observations as you're watching these things unfold and seeing what's happening from a different vantage point, not being necessarily in the thick of it but rather seeing it from the outside in and having conversations and starting to build a lot of relationships with faculty members, with department heads, with deans, program chairs, you name it, across the board, and just seeing how they're looking at stuff in different areas.

They all have great intentions. Of course, there is a small percentage of what I like to call "the curmudgeons" who exist in every ecosystem, and that became apparent to me early on, that you just need to avoid them. If you tried to change those folks, the wall in front of you would be full of dents made from your forehead, and you'd have all kinds of other issues. You could be handing out gold bars and they'd be like, "I like white gold," and you're like, "I have yellow gold. Sorry. Would you like this?" "Well, no, but I guess I'll take it if you're going to give it away…" No matter what it is, they're against everything that's happening in the system. That was another, I think, pivotal moment for us at the Kern Family Foundation and just in my own learnings and time. You don't need everybody to make change within an institution, and so that was one of the questions that we started asking, like, what is it you really need in order to change the design of an undergraduate engineering program at an institution? We started to see that we could impact 100% of students in a meaningful, robust ways, when about 30% of the faculty were engaged and bought in and moving forward in this area.

It's like, why spend your time on that 10% of people who are going to just wear you out, when you have these other people who really want to do this well and can connect and engage? For me then, I wanted to know, like in any system, who are the influencers? Who are the people who really have sway? And it's not a title. There are those people who, if they're involved in something and doing something, different people want to do what they're doing or understand what they're doing. Those were the people on the opposite side from the curmudgeons. I absolutely wanted to figure out who they were early on, because if they got involved and if they bought into what you were doing, their influence would start to permeate in really robust ways.

Then, I think the other part of the epiphany, I guess, was that change was resisted dramatically within higher ed, which is actually a real opportunity for us at the Kern Family Foundation, because we felt like if we could make the change, it's going to be hard to change it again. It was one of those things where you first looked at it as a challenge, but then it became an opportunity, because if you could start to bake this into the curriculum and the content, get people excited about it, and you could build that energy around it, it became a force to be reckoned with. You're not going to lose it in three months because one person leaves the institution, per se. It just becomes a part of who the department, the college, the institution is. It's like, you get the whole thing moving in the right direction, and like the Queen Mary, it's hard to turn the ship.

JW: Okay, so when you think about the lasting impact of KEEN and the Kern Family Foundation, do you believe the changes will persist? You've been able to change cultures, and you've been able to affect how engineering students are educated.

TM: I think so. I feel like there's been a leveling up within the expectations of what we want to see within our engineering colleges. It's not good enough just to have the skills to be a strong engineer. We need to have those attitudes, those motivations and dispositions that are going to continue to help students navigate the rest of their career, and that has become, I would argue, a common expectation now.

People are starting to wrap their arms around that. You need to have the skills so when you get where you're going, you can do the work that's in front of you, but if you don't have that mindset, you're never going to get there, or you're not going to progress in the way that you can. You're not going to have the impact that you can. You're not going to help create value for others in the way that you could, and that also promotes something that I think higher ed was always trying to promote, and maybe not always doing it in the most effective ways, which is lifelong learning.

When you graduate with your bachelor's degree, that's not supposed to be the end of your learning journey. That's the commencement into your journey of continued learning, and so now you need to build that into your design. Higher education was saying, look, we're going to give you the tools you need to be successful when you leave here, and now your job is to have the right mindset that's going to help you acquire the other ones that you need along this journey. And guess what, you're in the world of engineering in particular, and there is zero chance that we're going to give you what you need for the rest of your career when you graduate with your bachelor's degree. Zero chance. You're going to need other things in order to be successful throughout your career, unless your career is one to five years long, and then, in which case, we might have got you.

JW: Could you tell me a short story about one particular project or effort that captures all the things that we've been talking about, the importance of the communication for resistances to change? Is there a particular place that really stands out in your mind, a place that you think really gets it where they made it happen?

TM: The beauty is that you could probably point to any one of the institutions in the Network and tell that story, which is the perfect thing. That's also the thing that I like the most. Because of the things that we talked about earlier, and that sense of autonomy and every institution having its own culture and ethos, every journey was different, even when what they were shooting towards was the same. They had a common mission and vision around what engineering could be, but their journey, each of them was so different, whether it was this large public R1 or whether it was a small liberal arts college in a rural place. The way that they went about it was very, very different, but I think one of the consistent or common threads that we saw within each one of those stories was this beauty of (and I hate using the words "top" and "bottom"), but the top-down, bottom-up approach as connecting and really exceeding each other's expectations.

I think, for whatever reason, the leadership within universities doesn't feel like they can dictate what faculty teach, which is, ultimately, very true in the design of the institution, and so they would never even say, like, "Hey, this is something that's important," because they didn't want to come off that way. Academic freedom is important, yes, but that's not what you're doing when you emphasize what is important. You're just helping. If you think something's important, you can still, from a leader's perspective, say that it is important. You don't have to tell people how to do that.

The people who needed to figure out how to do it were the people that were actually doing it. I love the analogy of the ground troops and the air coverage. When you had that air coverage at an institution, it made your life so much easier. You knew that you had the support you needed, that the leaders were clearing the ground in front of you, that you weren't going to run into something that was completely unexpected because the academic leadership had a vantage point that they could see things that you couldn't from where you were as a faculty member. The opposite approach that most faculty take is almost like a covert operation. It's like they feel that they need to operate under the cover of darkness. They might think, hopefully no one sees me making these changes, and if they do, I'll just pretend like I'm not doing it for a moment. That way, they'll walk past and just keep going. And it's not like there's one institution where faculty feel that way. Every institution has it.

Part of it gets back to the communication. No one's ever said, "Hey, this is good. Do this," and maybe you got burned once upon a time as you were going through that program and so all of a sudden the assumption is, if I do that, I'm going to get burned or something's going to happen, so I can't do that anymore because you've been trained to not do whatever that thing was. That, I think, from a story perspective, you see that happening in so many different ways, but you can point back to the institutions that have really embraced EM and done it well, where there was that convergence between the administration at all levels, whether it be the president and the provost and the dean and the department chairs and program leads, with the faculty that were teaching first-year, second-year, third-year, and fourth-year courses, all coming together and really kind of seeing this blossom.

We also felt like that happened at a micro level, so it happened within programs and within departments, where you had the champions that were forming within the different programs, and then those programs started to really see the opportunities that were in front of them, embrace those opportunities, started doing things, and at the end of the day, those department chairs sitting back, going, "Wow, this is more than what I thought we were going to be able to do. This is really kind of powerful. The dean is thinking the same thing and they're seeing this."

I remember going to Saint Louis University, and President Pestello was one of the first people I got to meet when I joined the Foundation in 2015. He actually said something to me that I thought was basically the epitome of how you know it's working. I was talking about how SLU's doing some really cool things that we're really excited about and how that's making an impact in different institutions across the country, and we're basically just so proud of the work that's been happening in these different areas and some of these different people have been leading, so I thanked him for that.

His comment to me was, "That's great, and I really am happy that they're doing that, but more important to me is that they're an exemplar on my own campus. Parks College of Engineering is now the college that I'm pointing to in order to help everyone else see what they could and what they can be doing within their own colleges."

I think that's true. At the college level, you're seeing those different departments looking at engineering and saying, if it's happening over there, why can't it happen everywhere else? If it happened in biomedical engineering, if it's happening in civil engineering, if it's happening over here, this can happen everywhere. You do need institutions to buy in so they see it and believe it, until they can see it manifesting on their own campus and understanding the impact that that's it's going to have on their students and on their faculty. Once that happens, get out of the way because now it's viral, and it's going to transfer depending on who's involved in that department or in that program.

JW: If I'm capturing this right, EM is a catalyst for other change. It's not just going to happen. It's happening in engineering, but it's a way, a model for other colleges on the campus, other faculty to say, "Yeah, we can do this. It can happen." That's fantastic. I love that metaphor of the air support and ground cover. I love that one and I like that example from SLU, for sure. I'm going to ask two questions about advice, because the book is designed to help people who are not at a KEEN school, who don't have a lot of support, to find a community of change makers and connect with people that they could learn from. If you're going to give some advice to an academic change maker who might not be an engineer in education, but maybe any other place on their campus, what are couple pieces of advice that you might throw their way to help them along?

TM: I think number one is, find your people. Find your community, because, once again, I think sometimes we lose sight of the fact that we're all human, and we all crave that connection with other people who share some of the values and the things that we think are important. You ultimately have that in two ways within higher ed. You have that within your institution, but you also have that outside of your institution in your colleagues across the country. One of the things I didn't mention, but it was a huge epiphany to me in higher ed was, it is somewhat a lonely journey more so than others, partially because of the design, because you are bred to compete against one another, so even when you get your PhD, what do you do? You defend your dissertation, and it's all about you. It's you-centric. You have to prove that you did all the work. If you join other collaborators, you have to showcase that you did a substantial amount on this and you have to prove it. It's not a team effort, it's a "you" effort, and then you defend your contribution. All your evaluations are not about how you collaborate. They're about what you are doing. They're about how much money you raised for the institution, and how much money you brought in for research and scholarship. It's about how much you've published, and so, you're bred to compete against one another, and be even collegial enemies in some ways.

The people who you potentially want to hang out with the most are the people who have similar research interests as you. Those are the people that are also competing with you for the funding that is available and that you are going after, so over time I think you start to build some of that respect and you start to hang out with one another, but at the same time, there's always this healthy, weird tension around, "All right, I'm going to compete with Julia later because I want the funding that I know she's going to go after." There's this interesting dynamic, but, with that being said, if we could just get back to that collegial component of that work, at the end of the day, a rising tide raises all boats, that is so true. If we can start to work together more effectively and not focus on the "competition," but rather on the betterment of the organization that we're part of, about the betterment of the industry that we're part, the betterment of the research that we are driving, that is the goal.

When you do that, hopefully you are motivated by altruism more than necessarily anything else. You're trying to lean in and advance the research and the knowledge and understanding within certain areas, and so if we can get back to the root of what we're doing, I think that makes it a lot easier for us to start to form and build those relationships and to be genuine about it. If you can find those genuine people that are going to lean into this, and they realize that you're genuine in your actions, at that point, anything's possible. I think the problem is that there's always this lack of trust between certain parts of the organization or other things that are going on, and everyone feels like everyone's out to get them.

If we can get some of that out of the way, now all of a sudden you can start to really form some strong relationships, and those are going to be paramount to helping you get through the things that you want to do. Nothing worth doing is easy, so you're going to need some support and some assistance to accomplish whatever it is that you want to do. If your work is going to have any enduring effect and impact on large quantities of people, then you must find your people, find that community that can help you lean into, people who share similar passions around this work and who are willing to really work in lock step around doing something that's meaningful for others.

JW: Maybe likewise, what's the advice you should give yourself, looking back to 2017, what advice would you have given the younger Thor Misko at that point?

TM: That's a good question. Basically, I knew KEEN almost from its inception, and I would show up at some of the meetings, and I knew the Kern Family Foundation more or less from 2006, so that was when I started hanging out with the Kern Family Foundation. So what would I say to myself? I'd like to consider the advice I could have offered in 2015. I think, one of the pieces that's important is, there's that beauty in simplicity, and that's something I talk about way too much to begin with, and so I think, when I first came in, there was a couple things I saw almost immediately and then started to kind of forge, and one was, we were giving institutions grants in order to be a partner, and my thought was that almost backwards at that point.

At the end of the day, you should be a partner first before you're even eligible for a grant, and show, okay, so this is important to you, because if it's not important to you, then don't do it, but if it's important to you then let's do it. And if you have really good ideas, then let's talk about it and let's fund those ideas. This is a really simple idea, but it required a switch in the normal order of funding. It'simportant that someone is a partner before we fund, versus being a grantee before you're a partner. It's like, that doesn't make any sense to me. Maybe in the inception it was because you wanted people to try certain things, and so it was basically, you're buying their time to try some stuff and figure out if it's worth their energy, but at that point, there was so much energy around it already, it's like, this needs to change.

Then, I think, the other thing that was really important was that vision of entrepreneurial engineering simply being engineering. Just like you would never say, "Oh, is math part of engineering?" It's like, "Yeah, it is. You're going to do some math." This whole mindset and skill set thing is going to be part of what it is. It's not going to be called "entrepreneurial engineering," and I think there was some beauty in that simplicity of getting your message baked down to something that maybe doesn't encompass everything you're trying to do. It gets it to a point where people can look at it and be like, "Oh, yeah, okay, that makes sense, or I get behind that," or for the Network, it's like, "We don't want this to be called KEEN. We don't want it to be entrepreneurial engineering. We want it to be the engineering program. This is what engineering should look like," and so that changed the dynamic. We weren't talking about 10% of the students, just those majoring in engineering. We're talking about all students. This is something that every student should have access to, and this is what engineering should look like for all students moving forward. Creating some of those simple pieces earlier on, I think, was helpful, and I would argue that helping bring those outside perspectives in constantly and regularly to provide you feedback and help you see the things that are right in front of you that you won't see because it's the forest through the trees analogy. It's so true, though. We're so ingrained in what we're doing and we know so much about it because we're involved in it every single day, and someone on the outside says something and you're like, "What?" Usually, you get offended at first. You're like, "What? What are you talking about," but then you realize there's some truth in it and you're like, "Let me think about that for a second. Like, what?"

JW: Then you change your mind.

TM: Bring in not just your friends but those people who can help you, find your people, but find, also, some people who could be very candid with you about what they're seeing or what they're not seeing, so that you're not painting your own picture that isn't true. I think that took me longer to do than I wish it had.

I see now that it is part of casting a vision, and I understand that is important. As KEEN, we had to be able to paint a pretty strong picture with everyone around what we're trying to do, but there was also the reality of what was really happening in certain circumstances and those got neglected because we were kind of painting this imaginary picture of the future versus what was the current state. So you have to stay grounded in the current state and then have those kind of conversations so that you can help get people where they want to go and be that thought partner with them. I hope that is helpful.

JW: Yes, it's very helpful. So was there a question you thought I was going to ask you and I didn't ask you?

TM: The one thing that just came to mind, and I'm not sure if this is useful, but I was thinking about the air coverage thing. There's something that Rick Miller, president of Olin College, said once upon a time, which I thought was kind of funny, because we really spent a lot of time building relationships from presidents and provosts and deans, and then also across the board from the different areas, and I think that was critical. I looked at it as more of a way to connect people together that were already connected and help them understand that they're all trying to do things that are helping advance their larger goals, whatever those might be.

Rick's comment was, "I've never met a president who can actually get anything done, but they can sure make something **not** happen." The construct of the idea is, as a president, I can't dictate to my faculty what to do, but I can definitely tell them what they can't do. I thought this was another brilliant, simple thing, that you just pick up on. It's true when you think about it. Any leader at the end of the day can try to say "Do this, do that," but it doesn't necessarily work that way. You got to build that kind of excitement and you got to rally the troops, keeping this analogy alive, around the work that's in front of them and help them see the forest through the trees because we're all seeing the different parts of the elephant, all these analogies coming together, but you could definitely, like, at the end of the day, if you go over here, you're no longer part of our organization. You can set those rules up and start to create something that is or isn't, depending on your role, and your leadership.

I think the same thing is true with the dean of the college, so on and so forth, depending on the structure of the institution. That was another brilliant moment, which I think also led into why it was so important to have people at all different areas of the organization that were involved in this, in order to have sustainability for the changes that are made. Administrators won't be forever. For example, here in Milwaukee, I had lots of friends who worked as faculty at the University of Wisconsin Milwaukee, and they would basically say, "Well, we're not going to do anything that this administration wants because they're going to be gone in five years, anyways. We'll just kind of hunker down, do our thing, and then by the time they're trying to break us, they're going to leave. We'll break them before they break us." And that's not atypical in institutions. Because of the fairly consistent turnover rate in leadership in higher ed, it's easy just to kind of hunker down, lock into what you're doing, and don't pay attention to what's going on. Those two dynamics are important to understand. But academic change is bigger than any one individual, and if you can rally people at all different levels, all of a sudden it's going to endure because the faculty are the opposite. A lot of faculty stick around for a long time at their institution.

JW: Well, you say that you're an outsider in higher education, but I think the role that you played at KEEN was pretty much on the inside, and I appreciate you being willing to talk about it.

5 Propagating

INTRODUCTION

In her book *Making a Life: Working by Hand and Discovering the Life You Are Meant to Live*, author Melanie Falick presents rich portraits of weavers, metalsmiths, dyers, knitters, and other makers in order to show the "joy of making and the power it has to give our lives authenticity and meaning." The modern impulse to make derives, Falick says, from the specific conditions of our modern world:

> Over the course of just a couple of hundred years in the so-called developed world, we have become passive consumers of products, services, and information rather than active makers, fixers, and even thinkers. Most of the time, what we buy is made somewhere else, by a machine or people we'll never meet, sometimes working in conditions we would not accept for ourselves. Given these circumstances, it's not surprising that some of us are discomfited and feel a need for a grounding counterforce.
>
> (Falick 2019)

Paging through Falick's book, with its color photographs and detailed interviews with these artisans, you might feel an overwhelming urge to dig out the crochet hooks that your grandmother left to you and try your hand on a granny square as she taught you to do when you were eight. As a busy person, you probably won't have time to do any making, given that you are reading the introduction to this chapter in your five-minute break before your next meeting, after which you have to do the 20 other things that are demanding your attention any day of the week. But the impulse to make is as appropriate to academic change as it is to fiber art. In fact, as a change maker, you have been engaging in a kind of making all along as you crafted your academic change project and brought it to the Propagating stage (see Figure 5.1).

At this stage, consider what you are doing as less "making" and more "remaking," that is, what happens to your academic change idea when it is taken up by others and remade into a project that is based on your project but different—either in minor details or more substantially. This remaking proceeds in a way that reflects the culture and context of the remaker. For example, your academic change project may have focused on improving STEM education for a specific stakeholder population, such as women students in science fields at a research university. Then someone who hears about your project at a national conference takes up your idea and adapts it to a different stakeholder group who has similar needs but are educated in an alternative context, such as an institution that serves a different student type. When the adopter takes your idea and remakes it, two wonderful things can happen. First, the remaker has found a way to remedy challenges they see on their own campus but alters your idea to suit their local conditions and circumstances. Second, the remaking can inspire you to consider aspects of your project that may have never occurred to you before. The remaking is itself a chance for you to understand your project in a new way.

DOI: 10.1201/9781003349037-5

FIGURE 5.1 Propagating your change project. (Used with the permission of the author.)

In this chapter, propagation means the process of remaking that occurs when your original ideas are adapted by another. In general, as academics, we understand propagation quite differently. For faculty, the propagation of ideas is deeply connected to publication. This is how most academics envision it. You work on your project, collect your data, write up the results, publish an article or a book, then wait for the rest of the academic community to beat a path to your door. Your great ideas and innovative strategies are there in black and white for anyone to read and apply. How simple! Anyone who has taught an introductory course to first-year students, however, can see the problem with this approach. Yes, we require students to read the textbook (if we are assigning one) or the online course materials (if we have written and/or curated them), but consuming the ideas and knowledge isn't enough. Students need to apply this knowledge, work a problem set, write a draft, and run the lab, all with the goal of propagating the ideas into their minds and making them stick. If reading the published articles and reports about academic change happening in departments and on campuses were enough to propagate academic change, then we would have done it already. As Dr. Jeff Froyd elaborates in the interview for this chapter, change hasn't happened and won't happen this way. Instead, ideas about academic change propagate when you connect with people you wish to influence and persuade, using the tools and skills you have cultivated through this book thus far and demonstrating to them the advantage of transforming what they do and how they do it.

In this chapter, you'll add to your change maker toolkit by learning communication, teamwork, and diagnosing problems tools designed with a view to spreading your ideas and urging for their adoption by others.

- You'll find you are communicating even more than you did at the Initiating and Maturing stages but using different communication tools to get your message out.
- You'll need to enlist reinforcements for your innovative strategies, relying on the team development skills you cultivated earlier in this book.

- And despite all of your best efforts, you may confront the reality that your project is a failure. By diagnosing failure through the lens of failure analysis, however, you'll have a better chance of learning from that failure and apply what you learned to your next academic change venture.

You really didn't think I was going to let you give up, pack your tools, and go home, did you?

READ MORE ABOUT IT

Melanie Falick's book is a joy both to read and to skim through. Who knows, you might just find those crochet hooks and try your hand at making after all!

Falick, M. 2019. *Making a Life: Working by Hand and Discovering the Life You Are Meant to Live*. New York: Artisan.

STRUCK BY LIGHTNING (TALK)

Tool Type: Communication

Summary: At the propagation stage, change makers find themselves talking about their projects to a variety of stakeholders. A Lightning Talk is a great template that is adaptable to numerous audiences and situations.

In general, there are a few places you would rather not be when a thunderstorm rolls in: in a swimming pool, on the golf course holding a 9 iron, in an open field standing next to the tallest tree. But lightning in this section is all about the power and energy that comes from sharing your change project with a variety of stakeholders. At each stage of your change project, you need different communication tools, and the propagation stage is no different. What is different, however, is the purpose of communicating about your change project when you are working to propagate it with other departments or units on your campus, or with other colleges and universities, or with potential funders, or with governmental representatives, to name only a few. Suffice it to say that you are the best judge of who needs to hear about your project. By using the Lightning Talk form, you can decide who you can communicate with to the greatest advantage.

You may be curious how the Lightning Talk differs from the communication tools you learned in Chapters 3 and 4. The Lightning Talk offers a view of your change project from 30,000 feet. You and your team may find it difficult to rise up from the day-to-day details of your project, but not all audiences will wish to sit in the weeds with you. Instead, consider what your project looks like from that height.

As you will see from the template, the emphasis of a Lightning Talk is informing your stakeholders about the change project: the context in which the project operates, its purpose, and its impact. This is not to say that the Lightning Talk isn't persuasive. In fact, the skill and directness of an effective Lightning Talk can be very influential, as long as you adapt your Lightning Talk appropriately for the listeners.

- Have an opportunity to meet with a state legislator about your project? The base Lightning Talk can be customized to reflect the interests of this individual who may be searching for a new project to champion among her colleagues;
- Meeting with the new dean who just started their job in your college? Here's a wonderful opportunity to adapt the Lightning Talk to help this person understand the impact of your project and encourage them to support the ongoing work.
- Invited to be interviewed by an education news reporter about your project? The Lightning Talk is ready to be tweaked to suit a general audience.

As Rachel McCord Ellestad reflected during her interview at the end of Chapter 3, you may have occasion to present your Lightning Talk numerous times with new administrators, department chairs, and even university presidents. So let's get you ready to deliver some real thunderbolts!

Put the Tool to Work

A template is quite useful for getting a Lighting Talk drafted. Start first with establishing the context for your project. Consider explaining your academic setting so that your listener can visualize the environment for your project. For example, if you wish to write a Lightning Talk about the alumni mentoring program that you initiated and matured, then you might begin with describing the students the program serves, the needs these students have for mentoring that is provided by volunteer alumni from your college, and the needs identified by prospective employers and graduate schools who recruit students from your college. Since your project has evolved from the Initiating and Maturing stages and is now ready to be propagated, your objectives for the Lightning Talk have changed. You may be eager to propagate your project with other colleges who may be interested in adapting it for their own contexts. You could also pursue external funding to ensure that your project can be sustained through the propagating stage and beyond.

Now it's time to turn to your specific change project. Because the audiences for the Lightning Talk can be varied, list two potential audiences that you intend to communicate with.

You intend to share this Lighting Talk with:

Audience 1: who is this audience and why do they need to hear your talk?

Audience 2: who is this audience and why do they need to hear your talk?

Now you are ready to set the context for the talk. You will have the chance to customize the talk for each audience later in this section, so for now, the context you establish should serve for both audiences. Context in this case refers to the class, department, or campus environment in which your change project was implemented. Being specific about context helps your target audiences understand the applications of your project, rather than confusing them right off the bat. If you don't supply those details, their imaginations will fill them in and could send them down the wrong path.

Set the context:

```



```

Next, focus on the central purpose of your project. You are answering the question, "what do I want listeners to understand about my project by the end of 5 minutes?" For example, you may have a key message you wish to convey about the alumni mentoring program example used above: "As a result of their experience in the alumni mentoring program, our students have the opportunity to enhance their skills and expand their professional networks while they are still undergraduates." At this point, you'd like to customize the key message for each of the audiences who will hear this talk. If you are intending to share the talk with alumni at a different institution, you might sharpen the focus of the key message to reflect the alums' values and interests, like emphasizing that the program requires a manageable time commitment from the participants or it provides a new stream of workplace-ready students for their companies.

Key message:

```



```

In order to target Audience 1, you will emphasize this aspect of your change project:

```



```

In order to target Audience 2, you will emphasize this aspect of your change project:

```



```

Finally, end the Talk with a takeaway, such as answers to the questions: "What do you envision its impact to be?" "Why is it important?" "How will your department/college/campus be different as a result?" "What makes the project adaptable to different contexts?"

Key Takeaway:

```



```

Given what you know about the needs of Audience 1, your Key Takeaway for this audience is:

```



```

Given what you know about the needs of Audience 2, your Key Takeaway for this audience is:

```



```

The communication tools that you develop through each stage of this book need to be practiced and refined as you encounter new potential stakeholders and adopters. You might even consider carrying these versions of your talk on your phone, just in case you find yourself delayed at the airport and engaging in conversation with a potential adopter. It never hurts to be prepared and ready to create some lightning!

MY ACTION PLAN FOR THIS TOOL IS:

TELLING YOUR CHANGE STORY

TOOL TYPE: COMMUNICATION

Summary: Change makers can benefit from developing short descriptions of their projects that present a vivid and compelling picture of the project's impact.

Telling a story about an individual who is impacted by a change project is a powerful communication tool. By telling these stories to targeted audiences like students, faculty, administrators, community members, other stakeholders, change makers can share what is most meaningful and transformative about the work they are doing. You have already been introduced to the strategies of storytelling earlier in this book, and we are returning to it at this stage for a specific reason. Given that you are going to write your project story at the Propagating stage, your task as a storyteller is to tell a compelling story from the wrapping-up point of your project, rather than at the starting up or mid-point stage. And the best way to do that is to convey the impact of your project on an individual who has experienced specific benefits as a result. In order to ensure confidentiality, you don't need to give the real name of this person (pseudonyms are great!), but it is important to capture as much of their specific experience as it is appropriate to share.

PUT THE TOOL TO WORK

To start, let's focus on the end. If you consider many stories that engaged you as a child, they often ended with "And they lived happily ever after." I would caution against using the fairy tale ending in this context, but it will be helpful to have the final sentence of your story in draft, and that final sentence will help you focus the story you tell.

Let me share an example that could be illustrative in this context. When I collaborated with other Rose-Hulman colleagues to create a leadership development program on our campus, I captured the story of one particular student named Jason (not his real name) and told the story of his experience with participating in the program. He didn't see himself as a student with leadership potential, but I encouraged him to attend. I observed him and how he engaged with the program's activities, and it was from that observation and other conversations I had with him that I knew how his participation changed his view of himself and what he could achieve.

> **My final sentence for Jason's story**: Jason had a changed view of himself after the program was over. He was now the guy who could lead others on a senior design team, an identity that he had never considered before but now could see for himself and for his future.

The final sentence of your story is:

Now that you have a draft of a final sentence, work your way back through to the start. In my case, I would brainstorm about what the story content might be to capture Jason's experience with my project. I would use a couple of the program's primary activities, such as the team-building activity or the reflection writing Jason did. At this stage, I am keeping my focus on what I'd like my audience to take away when they hear my story. I'd like listeners to remember how my change project created a lasting impact for my student.

The element of your story you want listeners to remember is:

```

```

Now that you know where you are going and what you think should be memorable, it's time to do some brainstorming about what the story content might be. I suggest here that you start with some brainstorming that is targeted to the stage of your project.

Brainstorming 1—thinking back to the early stages of your project, what did you aspire to accomplish? What did you see as the potential for your project? What problem did your project address for a specific individual?

```

```

Brainstorming 2—considering what you now know about your project, what impact(s) were you able to discern then on a specific individual? Did you witness any moments of enlightenment in your project? Think of the story in the terms of a single person.

```

```

Brainstorming 3—now that your project is reaching its final stages, what has been transformative in your project for a specific individual student, faculty member, community member, or stakeholder?

As always, it helps to practice reading your story out loud to a friendly, yet critical, listener, before you attempt your story in front of a department head or dean, a committee meeting, or a funding review panel. You will be surprised, however, by how impactful your story will be.

MY ACTION PLAN FOR THIS TOOL IS:

CALLING IN REINFORCEMENTS

Tool Type: Teamwork

Summary: In order to propagate your project, you'll need to enlist the help of individuals with varying degrees of commitment to your change project.

When you live at the end of a gravel road 4 miles from a state highway and another 30 miles away from a town of any size, you learn the principles of self-reliance. You assemble a set of tools that can help you out in crises large and small, like a small rechargeable battery powerful enough to run the satellite internet modem when the power goes out, or a chainsaw that can cut up the tree that falls on the driveway during a storm. But there are situations and difficulties that require more than what a set of power tools can help you accomplish on your own. In these situations, you need a neighbor, especially a resourceful one, who can not only wield a second chainsaw but also owns a backhoe, drives a pickup truck, and can provide additional muscle when clearing tree trunks, wayward critters, and several feet of snow. Neighbors are, however, a source of help to be called upon judiciously. Self-reliance is the best policy day-to-day, but when help is needed, we can ask. When we do it, we do so with the understanding that the day may come when our help will be needed, and we will respond without hesitation. It defines being neighborly.

When a neighbor asks for our help, we respond positively. And when a work colleague asks for our help, our response is probably the same. We agree to help because it is in our nature as humans to want to help. It feels good to help. And perhaps in the back of our minds, we anticipate the day when we might have to make the same request. On the balance sheet of departmental obligations, we have built up some credit that we may need to cash in the future.

The sticking point is when you ask for help with your change project. Every change project needs help from others at different stages. While you can start the project on your own, eventually you will need to call in what Heidi Grant calls "reinforcements" (2018). These are the people you need help from because you can't move your project forward and achieve goals using your own talents and resources. Before you seek out reinforcements and ask for help, it's important to know what you need and how to ask for it. So that's why we'll take a moment first to identify **what** you need help with before you ask for that help, by using the Matrix of Needs. The Matrix is a tool that can you identify and prioritize where you need help the most.

Identifying needs may seem like an exploration into subjectivity. Ask a child what he needs after suffering a cut finger, and he might say "A bandage and a cookie." The first meets a physical need, the second an emotional one. The Matrix of Needs works on a different principle. Rather than identifying needs off the top of your head, the Matrix requires reflection on the state of the project (or considering where the project will be at a defined point in the future) and then identifying needs based on that reflection.

Put the Tool to Work

Let's start with a reflection on where the project is currently. Each topic is accompanied by some example prompts drawn from a variety of different change projects to help you focus your thoughts.

Expertise—As part of your new lab sequence, you have equipment that you haven't used before. Who on your campus, or among your colleagues located elsewhere, has the expertise you need?

What expertise do you need?

```
```

Organizational skills—Your project is at a level of complexity where tasks pile up and often fall off your plate. What you need is a project manager to help keep tasks and everyone working on them on schedule.

What organizational assistance do you need?

```
```

Staff—Your student recruitment project is exciting lots of interest from prospective students, but you don't have enough time to respond to every email request for additional information that you receive. You need a staff member to collect the contact information, add it to a spreadsheet, and reply to the student so they don't feel that you are ignoring their request.

How much staffing help do you need?

```
```

Financial management—You are a whiz with developing classroom pedagogy, but you have no clue how to use the college's financial management package to submit invoices and request checks. How do you learn this tool, or who could help you to complete these tasks?

What sources for help with finances can you call on?

```

```

Space—Your project is growing by leaps and bounds, which means you don't have the necessary space to serve the needs of stakeholders adequately.

Who manages space on your campus and what space could be reassigned to you and your project?

```

```

Time—While your work on the project has grown to fill your time, there is still more that needs to be done. It just can't be done by you.

Who might be willing to shift some of their time to your project?

```

```

Funding—Now that your project is showing clear signs of success, it's time to seek out sources of funding that can ensure its longevity.

What funding sources are available to you, both on campus and away from it?

```

```

The other aspect of identifying needs is using data to back up your claims. Can you show how frequently deadlines were missed, in order to demonstrate the need for a project manager? Can you tell a story about the impact your project has had on a particular stakeholder, thus making a persuasive claim for things you need, like space, staffing, and expertise?

The important thing to remember here is that these needs do not require a multi-million dollar grant or gift in order to justify the need. And the fact that these needs could be met with a request for help, rather than an effort by your Development Office, makes it more likely that you can find the help you need.

READ MORE ABOUT IT

Heidi Grant's book provides a rich resource on the research behind getting others to help you.

Grant, H. 2018. *Reinforcements: How to Get People to Help You.* Cambridge, MA: Harvard Business Review Press.

MY ACTION PLAN FOR THIS TOOL IS:

BUILD A FUTURE FOR YOUR CHANGE PROJECT

TOOL TYPE: TEAMWORK

Summary: As your project nears its end, you build a future for your change idea by identifying potential collaborators and adopters who will naturally wish to revise your project to suit their own contexts and goals.

In their book *Block Party: The Modern Quilting Bee: The Journey of 12 Women, 1 Blog, & 12 Improvisational Projects*, Alissa Haight Carlton, Kristen Leinieks, and Denyse Schmidt (2011) detail their work as they reinvent the traditional craft of communally produced quilts in a modern, virtual context. Taking up a model that our great-grandmothers were familiar with, Carlton and Leinieks assembled 12 quilters from around the United States and together they produced twelve quilts collaboratively. One aspect of their project stands out. While the members of the "bee" were each working on their own block that was ultimately assembled to make the finished quilt, they were also designing their block to reflect their own creative vision. Thus while each quilter was working with specific block dimensions, they were adapting the task to highlight their skills.

Throughout this book, we have focused on your change project and how you can best apply the tools of communication, teamwork, and diagnosing problems in order to ensure that your project becomes reality. At the propagating stage of your project, however, our focus is less on getting the project off the ground and operational than on sustaining it. The future may include continued support from your department or college, but you may also be interested in exploring how your project can be propagated with collaborators and adopters who are either on your campus or external to it. In order for others to adopt and adapt your project, however, you'll need to conduct two important assessments. First, if you haven't done so already, you'll need to identify the component parts of your project. These may be identical to the components you identified at the start of your project. These could also be completely new components that emerged through the evolution of your project from start to finish. Second, you'll need to spend some time identifying potential collaborators who may view your project as adaptable to their own contexts. In order to help illustrate this tool, I'll provide an extended example through an academic change project for which I serve as an external advisor. As you see the tool in action, consider how you can apply the principles to your own to propagate your project.

PUT THE TOOL INTO ACTION

The example project we'll examine is comprised of several components, some of which were described in the original project proposal and a particular component that evolved during the life of the project itself in response to an identified need. In order to build the future for their project, I asked the project team to identify and assess the components of their project. This is the first step in using this tool. With regard to your own change project, take a moment to inventory the project components, distinguishing between components that were part of the original concept and those that emerged as the project progressed:

Project Component Inventory: to capture both original and added project components

Project Component	Part of Original Project Concept	Added as the Project Evolved

I asked the project team to conduct the inventory in order to get them "out of the weeds," that is, out of the daily demands of their project, in order to understand what had been added to the project over the course of its life. For example, the original project proposal contained funding for student scholarships, an important component of the project that remained consistent throughout. Missing from the original proposal was a peer mentoring program. The need for such a program became clear to the project team members early on, so they developed the program and assessed it periodically through the project. As they neared the conclusion of their project, they recognized that the peer mentoring program was a significant component of their project and could possibly provide other collaborators and adopters with a model they could adapt to their own contexts. Thus, I encourage you to inventory all of the components of your project and determine how successful they have been, based on data collected during the project.

In addition, I asked the project team to reflect on who on their campus might find their project and its components interesting and possibly useful in their own contexts. If, for example, the project team found that the peer mentoring program provided important support to their students, is it possible that a different department or program might also see value for their own students? But who are these other potential collaborators and how would one find them? I suggested that the project team explore similar concerns as they may be expressed by other faculty, administrators, and/or departments on their campus. If we take the peer mentoring program as the illustrative example, then the project team can pose the question, who else is using peer mentoring to help support their students and increase their opportunities for success?

Collaboration Potential: to identify other departments/programs that are addressing a similar need

Possible Collaborator (either individual or the program)	Evidence of similar effort	Source of the evidence

By identifying similar efforts by individuals across your campus, you have a start on identifying potential collaborators and adopters. These individuals may be engaged in an initiative similar to your own. If so, then is there an opportunity to blend your efforts into a change project that can serve multiple colleges on your campus? If, as a result of your investigation, you find no other departments or programs addressing student needs in this particular, you may have found an opportunity to share your change project with others in order to encourage broader adoption.

In the case of the project team I worked with, they recognized that they would need to advocate for their project with their provost in order to rely on her leadership to connect and encourage multiple departments to coordinate a peer mentoring program. For that reason, you should consider how the positive relationships you have cultivated through the life of your project become even more valuable as you build the future for your change idea. I also reminded the project team that it was unlikely that any potential adopter would use the peer mentoring model they developed without first considering how to adapt it to their own departmental context. In this way, the project team who developed the peer mentoring concept and the collaborators in other areas of their campus were well on their way to creating their own academic change "bee," each contributor working with a set of specifications but everyone producing their unique program that reflects their own vision and creativity.

READ MORE ABOUT IT

Carlton, A.H., Leinieks, K, and Schmidt, D. *Block Party: The Modern Quilting Bee: The Journey of 12 Women, 1 Blog, & 12 Improvisational Projects.* Concord, CA: C&T Publishing, 2011.

MY ACTION PLAN FOR THIS TOOL IS:

RIGHT THIS WAY TO THE FAIL FEST!

TOOL TYPE: DIAGNOSING PROBLEMS

Summary: We don't often acknowledge the chance of failure in the change projects we initiate. In this section, you'll reflect on the role failure can play in your change work.

> But Mousie, thou art no thy-lane,
> In proving foresight may be vain:
> The best laid schemes o' Mice an' Men
> Gang aft agley,
> An' lea'e us nought but grief an' pain,
> For promis'd joy!
>
> "To a Mouse" by Robert Burns

So now you have read every chapter of this book carefully. You have written every reflection and completed each tool. You have planned, communicated, collaborated, and in general worked night and day to move your change project forward. At some point, however, it is clear to you that the project has failed, and you don't understand why. Perhaps the dean who supported your project and encouraged you to take the risk of pursuing change has left your institution to take a position somewhere else. Or the community partner who shared your vision for the work has been told to take a different direction. Maybe the circumstances for your project have changed on your campus. There are so many different possible reasons for a change project to fail that an investigation of failure could have been the focus of this entire book!

I don't wish to make light of your situation. In fact, I want to give you time to reconcile the plans you made for success with the realities of the failure because the lessons you learn from this failure will help you find the way into your next change project. Perhaps you thought I was going to encourage you to give up and go back to the ways things were before you decided to make changes happen for yourself and for others. As it turns out, I will argue for quite the opposite, in the form of a question:

> What does failure mean for academic types like us, people who have chosen a low-risk professional path and who may not be used to failure?

Not every experiment works. Not every design withstands rigorous testing. And it's because of the importance of learning from failure in STEM professions that we can analyze the possible sources of failure for the project. But before we do that, let's take a moment to acknowledge that failure can cause a severe blow to your confidence but it doesn't have to knock you out. Or to use the words of Scottish poet Robert Burns, the best-laid plans of mice and men often go awry, so let's figure out what we can learn from the failed plans to benefit us next time.

Recently I received a notice about the upcoming Fail Fest Wabash Valley, sponsored by our local economic development organization. During a one-day event, college students can brainstorm new ideas for marketable products, test those ideas, and perhaps fail, but then try again. The premise of the Fail Fest is expressed well in the advertising flyer: failure leads to insight, understanding, and innovation.

As I considered the allure of Fail Fest for undergraduate students, I was suddenly struck by a thought: where is the Fail Fest for college professors? Where is the safe, supportive environment that encourages creative, unconventional ideas? Where is the place where failure to accomplish a goal results in insight and understanding? Where is the place where failure results in moving a career forward?

It isn't easy to envision where the Faculty Fail Fest would take place since many of us in academic positions understand that failure in our workplace doesn't usually result in positive outcomes like insight and understanding.

- Try a new pedagogy? If it fails and students record their dissatisfaction in their end-of-term course evaluations, you may be tempted to turn away from innovation and return to the standard teaching approach.
- Explore an innovative research path? The tenure committee may not recognize or understand the work, or a journal editor may send your manuscript back unread.
- Put yourself forward for a new position in your department, college, or university? You may find that the new position goes to another candidate, or disappears entirely because of budget cuts and constraints.

As faculty, we are accustomed to succeeding. We graduated from college, we wrote dissertations, we were awarded PhDs, we sought and won research grants, and we were published. We don't often fail, and for many of us, it is very difficult to look upon any stumble or misstep as anything but failure. So, how do we prepare ourselves for failure and how do we celebrate the insights and understanding that come with it? Susan Robison, who consults with academics as a leadership coach and is the author of the *Peak Performing Professor*, recommends grieving the loss (2013). I agree, and her advice is relevant to the work we do as change agents. Loss and disappointed hopes are the price we pay when we take risks, challenge the status quo, and dedicate ourselves to change. Just because the plans we made didn't bear fruit doesn't mean that we won't see other positive outcomes at another time.

Let me suggest that you chart a path to an imaginary Fail Fest through a few important and intentional actions. First, reading books like Robison's and Jeffrey L. Buller's *Positive Academic Leadership* can provide context for what appears to be your failure. There is a noxious cloud around failure, and these authors can help you clear the air and emerge with a different attitude. Second, recall why it was that you put yourself forward in the first place: to change the status quo at your institution and to help create a new situation that benefits stakeholders. Third, I hope you can arrive at a point where you can emphasize the insight and understanding that comes from failure, rather than just the personal loss it entails.

So, welcome to the Fail Fest! You may not smell the cotton candy and the corn dogs, but at least you've bought your ticket and can see the Tilt-a-Whirl on the horizon!

READ MORE ABOUT IT

Susan Robison is a wonderful coach for academics, and she is wise in the ways of managing failure in any academic pursuit. You may find her book *The Peak Performing Professor* an important source of tools, information, and encouragement. Jeffrey Buller's book is also a source of positive inspiration (2013). I often find myself opening it when I feel that I need a pep talk from a trusted mentor and friend.

Buller, J. L. 2013. *Positive Academic Leadership: How To Stop Putting Out Fires and Start Making a Difference*. New York: Jossey-Bass Publishers.
Robison, S. 2013. *The Peak Performing Professor: A Practical Guide to Productivity and Happiness*. San Francisco, CA: Jossey-Bass.

MY ACTION PLAN FOR THIS TOOL IS:

ENGINEERING FOR FAILURE

Tool Type: Diagnosing Problems

Summary: By using the failure analysis tool commonly applied by engineers to failed designs, you can generate information that will help with your next change project.

When engineering students learn about failure, they often do so through a series of infamous disasters. In 1940, the Tacoma Narrows bridge suffered a catastrophic failure four months after it opened. There were, however, some early indications that the bridge would not function well in the setting; engineers on the project had already dubbed it "Galloping Gertie" for its propensity to undulate in even the most gentle cross-breeze. Similarly, the *Columbia* and *Challenger* space shuttles provide evidence of failure and are often used as case studies for poor data presentations. The *Deepwater Horizon* Oil Rig and the Fukushima nuclear plant failures offer their own powerful illustrations of failures that lead to environmental disasters. But faculty don't need to resort to such spectacular failures in order to make their point with students. A defective manufactured component, a faulty element, or an improperly produced device: each of these can help students understand the causes that contributed to each failure, and it is from those analyzing those failures that students apply what they have learned to their work in technical fields.

As the author of several books about engineering failure, Henry Petroski of Duke University encourages engineers to consider how important, even necessary, expecting and then analyzing failure is to moving technology forward. In response to an interviewer's question, "What is the value of failure?" Petroski responds:

> Success stories don't teach us anything but that they are successes. They are things to emulate, but the word "emulate" means two things. One, it means effectively to copy. Nobody wants to copy. Everybody wants to be more creative. They want to do something better. So "emulate" also implies trying to go beyond—trying to make it better, somehow bigger, whatever the measure is.
>
> Successes are not very interesting other than in that regard. When we do go beyond, then we move generally closer to failure. And what interests me about any failure is that it presents real lessons to be learned, because there's no ambiguity. When something fails, it failed.
>
> Generally, failure does several things. One, it shows us when something is not working as we had planned. That's one definition of failure. You design something, you expect it to behave or perform a certain way. If it doesn't, then there are lessons to be learned from that.
>
> (Used with the permission of the publisher)

While faculty may see the wisdom of introducing students to engineering failure, these same faculty may be less interested in applying the word "failure" to their own change projects. Rather than an opportunity for learning and advancing change, failure may be viewed as a door slamming shut on change, with associated feelings of personal failure and loss. It may be unimaginable to report to the funder of your project, yup, we failed and we even know why!

We can, however, take lessons regarding failure from the world of engineering and apply one particular failure analysis tool to our change projects. In doing so, we can learn important lessons that can open the door to further change efforts.

PUT THE TOOL TO WORK

The failure analysis tool that seems most useful in this context is the fishbone diagram developed by Kaoru Ishikawa (1976). There are plenty of tools that could be applied to understanding failed academic change projects, but the fishbone works well because of its simplicity and clarity. When you subject your failed change project to a fishbone analysis, you are working against your assumptions about what caused the failure. Instead, you are understanding the failure through possible causes and assessing the causes equitably, not assuming that one cause or another is primary. If we were going to analyze failure in a manufacturing plant, then we would use causes like "process," "material," and "machine." In order to apply the fishbone approach to a failed academic change project, we will need to devise a different set of possible cause categories.

In order to use a fishbone diagram to understand the failure of your change project, you will need to identify the causes that contributed to that failure. You can use the following fishbone diagram (see Figure 5.2), but feel free to draw your own fish with perhaps a friendly fish face. There's no need for your fish to be sullen.

First, at the head of the fish, define the failure. For example, you may have expected to recruit a high number of prospective students with the new major you and your team proposed, but response from students was lack luster. Each member of your team may think they know what caused the failure, but the point of using the fishbone is to conduct an analysis that can help identify causes that may not have been apparent. Thus, working back along the bones of the fish, each bone will serve as a cause that contributed to the failure. In the new major example project here, one factor might have been the marketing of the major to prospective students. If "Marketing" serves as one bone, then every aspect of the marketing effort can be listed along the bone, such as the fact that information about the major was not included in mailings from the Admissions Office and was never highlighted in the institution's social media. Other bones for this specific project could be the design of the major itself, faculty perceptions of the new major, and so on. The very process of identifying factors and related issues that contributed to those factors will provide you and your team a much clearer picture of what contributed to the failure.

Once you and your team have identified the categories of causes, each labeled on its own fishbone, then you can populate each cause with the factors that you associate with that cause, in the same manner. Another way to proceed is to brainstorm together all of the causes you and the team can think of without judging and throwing any out, then organize them into specific categories. Thus, you can build your fishbone from the outside in, or the inside out!

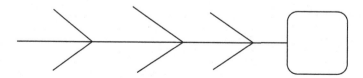

FIGURE 5.2 A rudimentary fishbone diagram made using Word drawing tools.

Once the fish has some flesh on the bones, then it's time to discuss together what the diagram has helped you see. In the new major example project, the team may have been under the assumption that the lack of a presence in social media was the cause of the failed major, but the diagram could reveal that a different cause was more influential, such as the way the major was presented during a face-to-face academic fair that new students attended in their first week on campus. While each fishbone diagram will reveal different aspects of each project, you may take away key lessons that will form the basis for your next project.

READ MORE ABOUT IT

You can read more about failure analysis using the fishbone tool by checking out the guru of failure.

Ishikawa, K. 1976. *Guide to Quality Control*. Tokyo: Asian Productivity Organization.

Of course, we have several academic researchers who have investigated failure from a less technical perspective. Adrianna Kezar (2012) provides useful insights in this article.

Kezar, A. 2012. Bottom-up/top-down leadership: Contradiction or hidden phenomenon. *The Journal of Higher Education*, 83(5): 725–760.

MY ACTION PLAN FOR THIS TOOL IS:

CHANGE MAKER INTERVIEW: DR. JEFF FROYD

I first met Dr. Jeff Froyd when I was a new faculty member at Rose-Hulman. He wasn't in my department, but I knew he was doing interesting change making across my campus. I volunteered to pitch in, and in the process, I received several useful lessons on change making. When I interviewed him for this book, Jeff was in the Department of Engineering Education at the Ohio State University after an extensive stay at Texas A&M University. Of all of the insightful and intriguing things Jeff shared during this interview, this idea sticks with me: "So you want this change broadly to occur, then you have to ask, how are the people and the processes that are involved going to be different in order to reach your goal?" Jeff has considered that question in his own change work, and I think it is useful for all of us to do the same.

Jeffrey E. Froyd was a Professor in the Department of Engineering Education in the College of Engineering at the Ohio State University. He received the BS degree in mathematics from Rose-Hulman Institute of Technology and the MS and PhD degrees in electrical engineering from the University of Minnesota, Minneapolis. He was an Assistant Professor, Associate Professor, and Professor of Electrical and Computer Engineering at Rose-Hulman Institute of Technology. At Rose-Hulman, he co-created the Integrated, First-Year Curriculum in Science, Engineering and Mathematics, which was recognized in 1997 with a Hesburgh Award Certificate of Excellence. He served as Project Director for a National Science Foundation (NSF) Engineering Education Coalition in which six institutions systematically renewed, assessed, and institutionalized innovative, integrated undergraduate engineering curricula. He authored over 90 journal articles and conference papers and offered over 30 workshops on faculty development, curricular change processes, curriculum redesign, and assessment. He served as a program co-chair for three Frontiers in Education Conferences and the general chair for the 2009 conference. He also served on the IEEE Curricula and Pedagogy Committee, which is part of the University Resources Committee, which is part of the Educational Activities Board. Prof. Froyd was an IEEE Fellow, an American Society for Engineering Education (ASEE) Fellow, an ABET Program Evaluator for the Engineering Accreditation Commission, a Senior Associate Editor for the *Journal of Engineering Education*, and an Associate Editor for the *International Journal of STEM Education*.

JF: My name is Jeff Froyd. I am currently the professor and chair in the Department of Engineering Education. I've been a chair for one year and two days. I came to Ohio State in 2017 as a professor in the Department of Engineering Education. I was recruited there because they told me they wanted someone with experience in engineering education who could work with the younger faculty members. The department had recruited three assistant professors and wanted someone who could work with them to help them get off to a good start. And they also wanted someone with experience in engineering education to help a new PhD program get off the ground (the department was created in November of 2015). So they hoped that with my experience I might be able to offer something useful.

JW: And before that you were at Texas A&M, where you were working on change projects on in a variety of areas.

JF: Yes, right.

JW: There seems to be a recurring theme in your professional life, being at different places and doing different kinds of change projects. So why are you doing this, why are you a guy who wants to take up change projects?

JF: I don't really want to take up change projects. But when I look at something… For example, the first project I got involved with was at Rose-Hulman. Within a year or two after I started there in 1981, I was teaching circuits. Circuits is the first electrical engineering course in the major, and students come into circuits thinking they're going to do electronics stuff. They think, we're going to design amplifiers and radio receivers and all sorts of fancy stuff. But when they get into the course, that's not what they do.

They learn circuit analysis techniques, which is so uninteresting, and there were no hands-on activities like building amplifiers. I just looked at the course and the way it was laid out and said, "Gee! Wouldn't our circuits and electronics instruction be better if we actually taught them together so that when student were learning circuit analysis techniques, they could relate them to electronics stuff, stuff that students wanted to do." That way, students could say "Oh yeah, I see, that's what I wanted to get into and now I see why the circuit stuff is in there."

I suggested to the department chair at the time that we revise the course, and he said, "Go for it." I wrote a proposal to the Lilly Endowment, and they funded me to sit down and sketch out what an integrated circuits and electronics course for sequence would look like, so I did. And it wasn't because I wanted to do a change project. I just thought this would be better for the students if they could see how the circuits and electronics work together. It was interesting because nine months or a year after we got started, there was a book that came out that arranged integrated circuits and electronics together. Now almost all circuits books have sections on operational amplifiers built into them because operational amplifiers are one of the easiest components to integrate into analysis. So it wasn't any kind of, "Oh yeah, I want to change this." I just thought it would be better if we did it this way. That's just how we did it.

JW: So this is a habit of seeing something that could be done better, and it has happened for you over and over and over again.

JF: Yes, but sometimes it bugs people. One of the things that I've learned is that when you talk to somebody about changing something, the messaging is incredibly difficult, because almost always when you talk to somebody about changing things, consciously or unconsciously, implicitly or explicitly, if you're not extremely careful, you manage to communicate to the people that you're talking to that what they're doing is wrong. And most people do not receive the information that they're wrong joyfully. When you talk about how this could be better, we could do this differently, we could make other improvements or whatever, the messaging is much more complex than I ever understood, especially early in my career. It's challenging to convey the need for improvement without simultaneously suggesting to people that they're doing something wrong.

JW: Did that get you cross ways with members of your department at that time?

JF: It got me crossways with faculty at the whole institution. I can remember talking to people particularly when we started the Integrated First Year curriculum, a program that integrated a number of subjects that are typically taught in first year engineering programs. We were talking to faculty about the ways a new approach could improve first year students' experience, and there were a couple of faculty in a meeting who said, "So you say we've been doing this wrong our entire lives?" Yes, okay, I didn't work that out right early on.

JW: So after a period of making these mistakes and realizing you were making them, you then changed your communication practice?

JF: Well, I saw the need to change my communication about change, but I still think it's really difficult. Nobody that I've come across has clearly articulated how you communicate a message for improvement that's not going to tell people that they're making mistakes or doing something wrong. So I still struggle with how to do that. Although I'm really aware of the need to do that, I don't think I have a formula that says, "Oh, here's how you convey that message." I do know, however, that it's possible to make mistakes in a number of ways.

JW: So I don't know if you could come up with a formula, because every context is different and every campus departmental culture is different. I think we should be aware and not just not waltz into the faculty meeting and expect everybody to say, "Hooray, you're here to set us straight and lift the veils from our eyes." Could you tell me a story about a time your communication really went well?

JF: I can't think of one project where communication went well, but because I know it's a challenge, I think I've done a better job of anticipating the challenge and trying to craft language that's going to come across more effectively.

JW: In the various projects that you have worked on, I assume that you've encountered resistance.

JF: I think one of the more difficult resistances to anticipate and counteract is the resistance that never bubbles to the surface, that's never brought to your attention. So in the years that the Integrated First Year program was running at Rose-Hulman, nobody ever came to my office, knocked on the door, and said, "What you're doing is bad or wrong." Instead, I would hear that there were conversations that were occurring between other faculty members, conversations that I wasn't a part of, and because it didn't come to the surface, I couldn't address it. If I were to say, "Dr. Smith, I heard that you said this," that isn't going to go anywhere. I think one of the more difficult forms of resistance to address is the kind where you don't have an opportunity to respond. How am I going to affect faculty conversations when I have no opportunity to influence the faculty conversation?

JW: Is that still a question in your mind: how to encourage those voices to come out, or how you're tuned into an environment enough to be able to say, "Oh, Professor X is not telling me to my face, but..."

JF: Well, the first thing I would say to people is, you should anticipate that this type of resistance is going to occur. Regardless of how well you've formed your message, regardless of how well you've tried to implement your plan, you should know that there's almost certainly going to be this undercurrent of questioning and resistance. So you should go forward thinking that this is actually happening. To the extent that you have an opportunity to address it in more public forums, you would not say, "Well, I know there's resistance going on, let me talk about it." But you should, I think, know that's going on and begin to guess what you might think is happening, what you might think the conversations are, and do your best to at least contribute some evidence that might redirect the conversation.

Two things that people conflate in their change projects are actually very, very separate: running an educational experiment or running a prototype. Someone may say, "We're going to run a pilot program." And because it's called a "pilot program," you can't figure out what they're doing. Some pilot programs are actually educational experiments. Those leading the project may say, "We're just going to do this, we're going to collect some data and we're going to see if there are positive results or negative results. We're just running an experiment." And because you're running an experiment, like you would run an experiment in the lab, you're going to get the results, but you're not really intending to change anything. You're just saying we're running the experiment, we're going to get some knowledge out of this and so forth. The second is when you're running a prototype program, and by that, I mean you're running the prototype to get some information on the implementation logistics, how it runs, and so forth, with the expectation that you're going to run a larger version of the program. We're not experimenting to get data, but we're running it so that we can see how it runs with a smaller group so that if you're going to mess up, you mess up with a smaller group and you get some information.

But the whole idea behind that is that you're running the smaller scale program to get information about running it at scale. And you would instrument those very differently. In an educational experiment, you would instrument two things to look at student learning, student outcomes, faculty outcomes, but you would instrument it to collect specific data. When you're doing the prototype program, you may collect data on outcomes, but you're more interested on how it runs, where were the weak points, will it scale up, what do we have to pay attention to.

I think there are a lot of pilot programs that are trying to do the two things simultaneously. And that's a problem. If you run an educational experiment, just like you would run any experiment, we do that in order to collect data and generate knowledge. We don't really necessarily want plan to take the results and immediately make any change.

JW: And if people were more cognizant of the difference between those two and understood if they're running the experiment, this is one path, but maybe not as threatening, but if they're going to run the other one, you better be forthright about what it is you're trying to do. You want to test the model, because you're planning to scale up into larger groups.

JF: That's right. And I think small scale programs would generate less resistance if they were clear that they are running an educational experiment. People could be very forthright and say, "We're running this program, we might run it for two, three years, but we're running it just to see what the results are. We're not planning to see this become the way we do things overall, we just think that it would be a very interesting thing to do." And it's a challenge because when you run this, you're doing this with human subjects and you're doing it over a timeframe, you can't run an educational experiment in 24 hours. And so you're going to run it for a period of time. But if you were very upfront that you're running the experiment, then people I think would be less resistant to it, because they see that it's not the case that it will turn into a bigger change. Whereas if we're running it as a pilot program, then I think you lay the groundwork for that very differently and are preparing people that this is going to be the way we do things in the future, and we're trying to collect data and feedback that will help us scale the program.

JW: So this leads me to my next question about your engagement with the Texas A&M RED project and your views of those RED projects that you've seen, both the ones that have been in existence since the start of the program and the newer ones. I don't know if you had any reflections about the RED program itself. The reason the RED program exists was to revolutionize engineering departments. It started out with this big aspiration, but I'm interested if you think RED is achieving that grand goal?

JF: The RED program is a successor to the departmental change program that NSF had earlier in the running around that 2002 to 2006 timeframe. I think they made the correct choice to focus on departments.

Engineering departments really are the place where, if you want to make change in engineering education, other than the first year or the capstone, then the department is the right focus. I think almost every RED project then proceeded to focus on the wrong things. And I say that because they will focus on things that they're going to do, like "We're going to introduce a new course, we're going to introduce some new program." The real change that needs to occur in engineering departments is in our faculty. We need to change how faculty think about whatever revolution you're trying to create. I think this is very similar to the initiatives in terms of diversity, equity and inclusion. Yes, we want the change, and so we run a program, but sometimes the programs do not focus on getting the faculty to change.

They focus on other things with the expectation that somehow faculty will change as a result of that. But really, you need to focus more on faculty development issues. People think about crafting programs first and then getting faculty "buy-in." I think this is a bad idea. For me, buying in suggests that you're a door to door salesman, and you're walking up to a faculty member (not ringing the doorbell) and saying, "I have some wonderful vacuum cleaner that I want to sell you." For an academic audience, faculty love to play the role of skeptics. So when you say, "I want faculty buy-in," you're immediately taking the faculty and saying, "I want to put you in the role of the skeptic and then I want to try to sell you something."

In the 2000s, we wrote a paper that didn't get as much notice as I thought it was going to do, but at the end of the paper, we said that after 10 years of work in the Foundation Coalition, when we surveyed faculty who played crucial roles in their curriculum development projects at the institution, it was never about curriculum change; it was about faculty change or faculty development.

JW: Have you seen any programs that seem to be taking or adopting a strategy more along the lines of what you're talking about, a faculty development strategy rather than a program development strategy?

JF: There are a lot of RED programs that I'm not intimately familiar with, but when the RED programs would gather for an annual meeting and people would talk about what they're doing, you'd rarely see that their focus was on faculty development. They would talk about the new course they're trying to put in place or the new program they're trying to put in place. They would rarely talk about how they want the faculty to think and behave differently.

JW: That's a great phrase: think and behave differently. I had a question. You've done a lot of work with dissemination and propagation. I know that when I talked to you at RED meetings, your concern is with how RED projects share their work with other departments. There is a general expectation that people think, "Well, I'm just going to share this with the world," but there is a huge step up for other departments, other people to look at what you did, look at what I did at Rose-Hulman and say, "Oh yeah, we're going to adopt that," because they can have that whole laundry list of reasons why it won't work on their campus. Could you talk a bit about dissemination and propagation and why that has become an important concern for you, something that you've been devoting quite a bit of time to?

JF: Why is it a concern for me? Well, I think NSF has invested literally tens if not hundreds of millions of dollars in projects where their espoused intent is to fund projects in order to influence how people do things at other campuses. There's just lots of projects where NSF has said that they are funding this so that it will be done here and they expect that their funding at this point will result in change at other campuses to some degree. I think this has not occurred to the extent that NSF may have expected, but I think NSF believed that this kind of program will get funded and the PIs will publish their results and people will look at it and go, "Yeah, we want to do this." And that's what I regard as the dissemination approach in people. NSF would say, "We want to know how you're going to disseminate it." And people would talk about how they're going to publish papers and present at conferences and maybe even do a workshop. But the focus was on just disseminating the material, getting it out there. The PIs believe that if they get it out there, then others will adopt it, because they have seen that happen in other scientific disciplines. If I have a new way to treat a particular disease and I publish that the results are effective, then they have seen that other people respond.

I just don't think that works because we are talking more about human change when it comes to educational projects. I don't think the dissemination approach works. And I think that they need a new vocabulary and a new, to use an early to late 1960s word, "paradigm," about what projects NSF funds and how that funding is going to influence other people. During a proposal review session, you can read many of the projects that were submitted to NSF and the language in the proposal emphasizes the uniqueness of the campus and the uniqueness of the program and how the uniqueness of the program fits with the particular situation on the campus. And PIs explain, "We are the perfect place to run this particular program." And I think, "Great, then go get funding from your own institution." Why would the National Science Foundation fund a program that is so particular to a campus when the very nature of that suggests that nobody else is going to want this?

JW: I've never heard someone describe this situation using a scientific disciplines analogy, that its uniqueness is absolutely the thing that is most likely to jeopardize its impacting somebody else and their likelihood of adopting it. So if there were an alternative then to the practice of funding uniqueness and having things just show up in a JEE article or at a workshop, what's the alternative?

JF: I've seen some proposals in which the PIs have gathered people from two or three different institutions, maybe where the institutions differ in some characteristics, and they have agreed to work together to look at the aspects of the program and how it works differently in different campuses with different cultures and with different faculty. They have already established a relationships with people who have indicated interest in even trying this. They aren't yet committed to do it, but they are exploring it and are interested to see how it works on the different campuses.

A proposal like this is already anticipating that it needs to adapt to different kinds of campuses, to different kinds and sizes of programs. They already are anticipating that in order to see it go to other cases, they've reached out and said, "Oh, well, we have some people who are interested maybe on an advisory board or whatever, who can observe the results on a more intimate basis and be familiar with some things that transpired," and they could even provide some input regarding how their particular campus would need to consider changes in the program. But that to me is an example of something that's more likely to propagate than something that emphasizes the very distinctive nature of the campus and the program. But even within the last two or three years, despite NSF increased emphasis on seeing their funding have more of a broader impact, reviewers get very excited by novelty.

JW: I also wanted to ask you about the kinds of change you think you're affecting at Ohio State where you are currently. You're a department chair, you have administrative power, you have named power, so do you think this is the only way to make change happen? I don't believe it is, but I wanted to hear what you had to say about it.

JF: If I had to point to something that I think has had the largest influence on our department, it would not be through any seniority or named power. It's more about the work that I've done with the younger faculty members in meeting with them and talking about a draft of a publication or proposal that they intend to submit. I would try to provide some feedback on the publication or proposal and talk with them about it. I think efforts of this type and working with the faculty members on those types of projects has been much more influential in terms of the development of the department than any named power or anything that flows from either seniority as a professor or chair title.

JW: It probably points to you and other members of your department having a very different view of who you're going to hire and how different parts of their careers are going to be encouraged. I think this is a big part of the faculty development work that you do and I think good chairs should do. We will keep repeating over and over the same resistances to change if we keep hiring the same way, rewarding the same way, and developing the same way.

JF: I agree that it's the investment in people either in terms of the work that you do to hire faculty members better which will help the department become more productive or the extent to which you invest in their development. One of the very influential things we talk about with respect to younger faculty is the role of mentoring. Mentoring programs have, for me, been a real sore spot because the assumption is when a department has a mentoring program, a new faculty member will come into the department and benefit from all of our combined wisdom. The department, not the new faculty member, will pick the exact person who they believe is going to be the perfect mentor for that person.

And because the department has picked one person and said that they're going to work with the younger faculty member, somehow, we have now created the perfect faculty development program, because that one mentor is going to do everything. This was brought home to me by the model adopted by the National Center for Faculty Development and Diversity. They have one of the best models I've ever seen. They place the new faculty member at the center of the circle. Branching out from there is a set of numerous spokes, and each of the spokes is a need of the faculty member: career advice, help in teaching, help in relating to other faculty members in the department, and so on.

It's clear that no one person can satisfy all those needs as a mentor. We need to be more intentional about understanding the needs, then helping the faculty member identify resources that they think are going to be helpful to them rather than to believe that we have the wisdom to pick the right person for them. For example, some programs have created a faculty development in which they say, "Young faculty members need to know more about their field and about what are pressing problems in the field."

And so they will say to the faculty member, "We are willing to support you in your effort to define an expert in your field who you believe can help you, and we will then marshal resources to make it likely that this person will help you and maybe even make the ask, but you have to tell us who this person is." That seems like a better way to go at it, since the approach recognizes that one mentor can't meet all of your needs. We just know that you need to have a better picture of your field. That's one of your needs. And we think we're going to help you find a way to fill that need. And then we will neglect other needs or whatever. But I think too often we have programs that are based on our supposed wisdom in picking people who are going to really help faculty members with all of their needs all at the same time. And this is something that the RED programs could have done. They would say, "Okay, we are going to need our faculty to think and behave differently, who can we get to mentor them?" I've never seen a program like that.

JW: If you were going to share some advice to others who are working on academic change projects, maybe either at the department head level or at the beginning of their careers, what would you offer?

JF: I would say, okay, so you want this change broadly to occur, then you have to ask, how are the people and the processes that are involved going to be different in order to reach your goal? Too often a program says, "We need this and so we're going to implement a program that does that." Okay, great. But if we want it to get there, what are all the things that are going to act differently, whether it's the individual faculty members, whether it's some of the institutional department or institutional processes that need to work? I think of the integrated curriculum we talked about at Rose-Hulman, and it worked across four or five, six different departments, and never in our thoughts that we ever go, "Huh! If this is going to work across the departments, what institutional processes are we going to have to put in place that are going to facilitate this working across departments or whatever."

We just said, "Well, we need a curriculum, put the curriculum in place, teach the curriculum." We never got to the point of ever discussing, what are the institutional processes that are going to actually have to be different in order for this to go forward? In general people do not invest significant efforts in examining what do we want processes and people to do differently that are going to enable us to reach our intent? And if that is what we need, if that's how we want people to behave or processes to behave, what are we going to do to get that change put in place?" That's the biggest piece of advice I would offer.

JW: I'm going to give you an opportunity to tell me what question you thought I was going to ask that I haven't asked and that you really wanted to answer.

JF: So we talked earlier about how focusing on talking to people that we need to improve, manages to convince people that they're doing something wrong. That made me think of a TED talk, it's called "On Being Wrong". And the presenter Kathryn Schulz actually wrote a book with that title.

In the talk, she asks people in the audience, what does it feel like to be wrong? And you can see the faces of the audience, and they each have pained expressions. Then she says, "You're answering the wrong question. What you have answered is what does it feel like when you find out that you were wrong? What I'm asking is, how does it feel to be wrong?" And she then says, "It feels very much like being right."

And she relates a story about a patient brought in for surgery on their left leg. The surgeon performs the operation, but when the patient wakes up, they ask "Why do I have all of these dressings on my right leg? It was my left leg that needed fixing." And they ask the surgeon. The surgeon says, "All the time that I was doing the operation, I thought I was working on the correct leg." What does it feel like to be wrong? It feels very much like being right. We don't really think that when you're talking to people about academic change and they're wrong, maybe in a situation, they feel like they're right, because feeling when you're wrong is feeling like when you're right.

So I guess one of the things that I've noticed about myself is that I have a greater unwillingness to initiate change projects and go, "Oh, I see a way to make that better. Okay. We've done this many times, Jeff. Let's take a real serious look at this and begin to prioritize and ask, Is this really something that you want to spend a considerable amount of time and energy investing in?" I think people routinely underestimate the time and energy it's going to require to make the change that they have in mind. And if they were more aware of the resources that they're going to expand, they might be more measured in terms of their willingness and exuberance to undertake the project.

JW: But is there also an opportunity there? You have identified a need, but it may not be your project, your time to work on it. This could be a project that a person coming up behind you might want to work on, right? Doesn't always have to be you, Jeff, or me, Julia, saying, "Oh, we're going to go to the ramparts on this one." But I think that points toward recognizing that in order to develop junior faculty, there's a moment of saying, "Hey, this may not be my time for this project, but it could be yours," or it could be you taking the idea and running with it in totally a direction I didn't even think about going. So I think recognition is a skill. This is a real skill.

JF: I participated in the initial version of the I-Corps for Learning project that NSF put in place. They were pleased with the results of their I-Corps program. And so they thought they would run I-Corps for learning for their educational projects. And one day they brought in some people externally and one of the people said something that has stuck with me, and I think it is almost the exact opposite of the way faculty members tend to approach things. When you see something that needs to be done as part of your effort, many faculty assume they need to add it their already too long to-do list.

But it is clear that non-academics view this very differently. Their first thought is, this needs to be done, who can help me do this?

6 Arrivals and Departures

INTRODUCTION

Just beyond the security gate, in the main lobby of the airport, you can see them: the family, the friends, the spouses, and all those who are waiting to greet the new arrivals (see Figure 6.1). No matter how long they have had to wait because of bad weather or poor flight schedules, they are happy when they see the people they have waited for. The greeters bring flowers and hold signs. They wave and cheer. They offer hugs and kisses. They have anticipated your arrival, and they make the travel worth it. They tell you in their words and actions that you have arrived, and they are so glad you are here!

It would be wonderful if there were greeters who could welcome you as you arrive at the end of your change project. They could hold signs made specifically for you: Great Job! You Made It! We're So Proud! The greeters would be the indicator that you made it to the end of your change journey. By now, however, you probably have figured out that there will be no greeters to welcome you upon arrival. It's likely that you won't even realize that you are at the end, and the end may catch you unaware. Your change project may have a specific ending, like the end of the funding or the

FIGURE 6.1   Arrivals and departures and your change project. (Used with permission of the author.)

DOI: 10.1201/9781003349037-6

submission of a final report, but in many instances, the end will be unclear. The end might even be signaled by the beginning of the next project. Whatever your case may be, take a moment now to greet yourself as you arrive at the "end" of your project. Great Job! You Made It! We're Proud of You!

Rather than spend time trying to determine definitively where the end is, let's focus instead on the three categories of change tools that have organized this book and what the end means for your communication, for your team, and for your ability to diagnose problems. Now is the time to reflect on the journey you have made.

In Chapter 2, you created a Roadmap for Change, a graphical representation of the project you planned to work on, the challenges you expected to face, and the destination where you hoped to arrive. If you haven't reviewed your Roadmap for a while, this is the time to unearth it from under the pile of papers on your desk and consider it from the perspective of a change maker who has arrived, instead of the change maker who is just starting out. And rather than overwhelm you with questions about your project and your experiences, your ups and your downs, you can best serve yourself by considering one question:

If you had the chance to do your project all over again, what would you do differently?

Take a moment and write reflectively about this question. It may be that the thing you would do differently is regularly schedule time for you and your team to meet and assess the shared vision you have for the project. Or you may believe that you needed to pay more attention to how you communicated with external stakeholders. Every change project is different, and every change maker finds different strengths and weaknesses that emerge from the project context.

Based on your reflection, complete the following sentence: If I started my change project over again, I would _____

Rather than compose a long list of do-overs, focus on the one thing you could do differently that would improve your project, increase its impact, or advance the development of you and your team. By isolating what you would do differently, you have the chance to create a new Roadmap for your next change project, since the end of one change project doesn't signal the end of your career as a change maker. The end of one project means the start of your next. Your arrival is, in other words, just a departure on your next change journey.

CHANGE MAKER'S APPRAISAL

Summary: You may be familiar with self-appraisal as part of your annual evaluation by a department head or dean. A modified appraisal form can be useful as a tool to help you identify your accomplishments as a change maker and areas where you may need further development.

Make no mistake: the world loves a rating event. It could be an episode of a national singing talent show or the compulsories of an Olympic figure skating competition. No matter what the sport or musical genre or unusual skill, we love to see the judges affix those scores (and also argue with their ratings!). How different would these events be if, instead of the judges giving a numerical rating, each contestant would self-appraise? The figure skater would glide back to the bench and take a moment to review their performance. Then, directly addressing the television cameras, they could rate themselves, highlighting what went well in their performance and where they needed additional practice. There could even be a chance at complete magnanimity, when the skater identifies the skills of another competitor, with a comment like "I thought Adrian totally nailed that triple salchow! They deserve 10 points! What a wonderful change this self-appraisal would make in our experience of the competition!

You may be among the critics who look at self-appraisal with suspicion. You may point out that athletes could overestimate their achievements and wouldn't honestly evaluate themselves. The opposite is more likely. Many athletes, performers, and some students are quite self-critical. How else can they identify areas of weakness and work to improve them? You might also suggest that by allowing contestants to rate themselves, we would lose the objectivity that judges provide to these competitions. The Olympics, however, have had difficulties with ensuring objectivity among those conferring ratings. After all, we humans have a difficult time setting our personal views aside.

Of course, there is inherent difficulty with conducting self-appraisal and making it meaningful. For example, some years ago, I received a copy of the self-appraisal form used by a Fortune 500 company that hired many of my college's graduates. The form was emailed to me by a senior student who was interning at the company, and the form had been used by engineering managers with summer interns as a way to introduce them to the idea of employee self-evaluation as part of an annual review. Today such self-appraisals are part and parcel of modern workplaces, but at the time it wasn't something customary practice. Because I saw the application of the self-appraisal practice to my undergraduate students' future careers, I introduced the form to them and used it as part of the teamwork component of their class projects. Each student appraised their own work and contribution to the team project, then those self-appraisals were compared to the assessments each team member completed of the other team members. In principle, it seemed like a good idea.

Except it failed with my students. When asked to appraise themselves and their work on their team projects, students were reluctant to review their contributions and identify their strengths and weaknesses. They preferred to wait on my judgment since I was the one who was assigning a grade for the project after all. Some students were overly critical of the work they did, minimizing their contributions and deferring instead to other team members. Others tended to inflate their contributions, using the self-appraisal as an opportunity to toot their own horn and to make up for less-than-stellar performances. As I reconsidered using the self-appraisal form with my students, I recognized an important aspect of the tool: I couldn't expect my students to conduct accurate self-appraisals if they had received no coaching on how to do it. The skills associated with self-appraisal—the ability to evaluate their own work accurately, the ability to identify strengths and weaknesses, and so on—was not something they had been asked to do throughout their educational careers, and so it was up to me to cultivate these skills throughout my course. Once I had that realization, the subsequent uses of the form and the students' self-appraisals were more successful and satisfying.

From skaters to engineering students to change makers, we all need to cultivate the skills of self-appraisal. And as you near the end of your change project, you need to take a moment to assess what you achieved in your project and how you added to your change maker toolkit. With these insights identified and documented, you will have a launching pad for your next change effort.

Put the Tool to Work

In the Change Maker's Appraisal, you will evaluate both the project you implemented and yourself. Your evaluation of the project can help you in several ways. You will be documenting your change project, which can have important benefits. For example, you may be asked to take on a new project based on what you were able to achieve in your current project. Or you may need to explain the impact of the current project in order to secure funding for the project's next phase. Either way, evaluating the project you are completing is useful. The appraisal helps you understand how you have grown and developed as a change maker, what tools you found beneficial, what tools you probably should have used but didn't, and what you will need to pay attention to in your next project.

Start your appraisal with some brainstorming:

List three major accomplishments that you achieved through this project.

Record the major publications that were produced during this project.

Identify the most important collaborations that emerged during the project.

Based on what you were able to achieve during this project, identify three goals that you would like to pursue with your next change project.

Looking back over the life of your project, identify where you made a difference on/ for a stakeholder, a program, a context, etc.

Now turn your focus from your project to yourself, the change maker:

As a result of this project, the three most
important changes you made to your
communication were...

As a result of this project, the three most
important changes you made to your
teamwork were...

Identify the new skills you gained as a
result of this project.

What outcomes from the project are you
the proudest of?

What outcomes do you wish you had been
able to achieve from the project?

In your next project, what three aspects do
you plan to pay more attention to?

Remember that self-reflection at the end of a project is built on the skills you developed through the life of the project. If you have been using this book as part of your change making effort, then you have had numerous opportunities to reflect on each tool you were introduced to. And if you are, like some readers who take up this book, the person who skipped the reflection prompts along the way, it's never too late to make up for those missed opportunities. At any rate, you can ensure that you understand what new skills you have acquired and the skills you still need to add to your toolkit if you make time to evaluate both your project and yourself as you near the end.

A VISION OF YOUR FUTURE(S)

Summary: The change maker's work is often focused on improving a current condition in order to affect the future. In this section, you'll consider a possible future in which the work you have done in the present has manifested the changes you sought to accomplish.

It's 2032, and you have been invited to give the plenary presentation for the International Change Symposium, a gathering of change leaders and makers who share your dedication to transforming academic contexts. What will you tell this rapt audience about your project as it exists ten years from today?

You start your preparation for your presentation with some brainstorming:

1. What elements/features/components of your project will exist in the year 2032?

2. What elements will need to be changed? Why? How?

3. What partnerships can you envision for your projects in the future?

4. How has your project impacted your department, your campus, and higher education?

In preparation for your plenary, you need to get the word out via social media. TikTok videos are the primary forum for sharing news in 2032, so you and your team need to create a two-minute video that will capture the major aspects of your project based on your brainstorming work. Use the space below to draft your video script and determine who will appear in the video. So why not record your two-minute video and share it with the world?

CHANGE MAKER INTERVIEW: DR. KHAIRIYAH MOHD YUSOF

The interview with Dr. Khairiyah Mohd Yusof concludes the change maker interviews for this book because her experience as a change maker spans a long and diverse career. During our conversation, she reflected on the aspects of her change work that align with her training as an engineer. Her problem-solving perspective has inspired much of her efforts to make change.

I was also struck by her advice to change makers:

> We need to lead. We need to lead by example. Have patience... As a Muslim, we believe that you cannot change other people's hearts. Only God can change other people's hearts. So you do your best, but then God will decide whether it will work or not, so it's okay.

Dr. Mohd Yusof is the founding director of the Universiti Teknologi Malaysia Centre for Engineering Education, promoting meaningful research and scholarly practices in engineering education. She is on the editorial boards of *ASEAN Journal of Engineering Education, Journal of Engineering Education, IChemE Journal of Education for Chemical Engineering, European Journal of Engineering Education,* and the *Journal of Problem Based Learning in Higher Education.* She is currently on the executive committee for the International Federation of Engineering Education Societies (IFEES), President of the Society for Engineering Education Malaysia (SEEM), and a board member of the International Research in Engineering Education Network (REEN).

Khairiyah Mohd Yusof (KMY): My name is Khairiyah Mohd Yusof. I'm an academic staff at the Department of Chemical Engineering, Universiti Teknologi Malaysia. I was the head of the Chemical Engineering Department way back in 2005, 2006. Then after the stint as the head of department, my term was renewed, but then I was pulled to the administration to be the deputy director of the Centre for Teaching and Learning in charge of faculty development. So just after my first term as Centre Director, I was changed again to become the director of the Centre for Engineering Education, which was just formed. I was the founding director, and that's where I was until a couple of years ago when I stepped down, so now I'm waiting for my retirement next year.

Julia Williams (JW): So in these different roles that you've had, you've been trying to make change at UTM, not just in chemical engineering but all across the School of Engineering and trying to have that impact. So my question is, what motivates you to pursue change projects? Why are you a change maker?

KMY: Well, change, for me, is normal. If you don't change, you get into a rut. You are in your comfort zone, and you don't really grow. And the world is changing. If you don't adapt to what is happening in the world, we'll be left behind, and nothing will work.

As an engineer, you are supposed to solve problems, and engineers are always very pragmatic. You know what is the ideal, but then you know you cannot reach the ideal because the world is complicated. So you get to the best that you can, given the current availability of the resources and whatever constraints you have at that time. So when I was teaching my students and they were failing, I viewed this as a problem, even though it was not an engineering problem, but I'm an engineer, so I conceptualize it as a problem. I don't want 30 to 40% of my students to fail. I was talking to other faculty members because our program is quite big. We have three or four parallel sections teaching the same subject. I was not happy that my students were failing. In talking to other faculty members, they said, "It's okay. That is normal if you have three sections of the same course, and you have 30 to 40% failures." So we had to add another section just for the failing students. That was a joke. I said, "This cannot go on."

At the time, I was studying for my master's and my PhD, and I had been interested in at least reading about engineering education, reading some journals, getting newsletters like Tomorrow's Professor just summaries of the current research in teaching and learning in higher education. So I couldn't accept the fact that 30 to 40% of my students will fail, and very few students earned A's. So I decided to change the way I teach.

All of the faculty teaching in parallel sections were giving lectures. Obviously, that didn't work. So I experimented. I knew from my reading that faculty in other parts of the world were also doing this. So even though those around me were not doing it, I said, "Why not? You just try." So I did, and amazing things happened. But then once I did that, I wanted to get others to also try it. But being in engineering, you don't really know how to approach things. I needed to identify who would be willing to try with me, and who would be against these ideas. I tend to avoid difficult conversations because it can be quite stressful. And sometimes faculty in engineering are not the most pleasant people to disagree with.

So it was not easy. But then when I was head of the department, I managed to use whatever power that I had to persuade people to do things. So the younger faculty members were easier to persuade.

JW: And when you encountered resistance from colleagues, what forms did those resistances take? Did you hear particular expressions about why you can't make the change, why it won't work?

KMY: Well, the culture in Malaysia is very much teacher-centered. We don't have to do student-centered learning. We don't have to do active learning. We've always gone through lectures.

JW: You said you had success with junior faculty. Was there anything you learned about persuading people among the junior faculty that you're able to apply other places? Is it merely the fact that they have no choice because they want to be tenured or kept on in contracts?

KMY: No, I was also learning. We invited experts from other universities, and I encouraged junior faculty to go through the training that I went through, and when we wanted to implement things, we would discuss how to do it.

At the same time, our university started taking an interest in engineering education and came up with a conference on engineering education. So because we were doing something different, we had something to write about in scholarly papers.

I had always read engineering education papers. So for me, it's natural to be writing about it just as I would with the technical papers. Even before UTM had a conference, I had written two or three papers before that. So then I told my junior colleagues, "Come on. Let's write this together, and each of us will head one paper. So this person will write about this. This person will write about this. We'll write about something." In this way, we could all contribute to each other's papers so we would end up with four or five papers at the regional engineering education conference we hosted at UTM.

At the time I didn't have the money to invite the big names like Richard Felder and Rebecca Brent, Karl Smith and other notable scholars. By holding conferences, however, we could actually bring the experts in. We found the money on how to cover the cost. The purpose of the conference was to draw everybody into engineering education. We can get to know others who are really interested in engineering education, but at the same time, I invited experts who I wanted to learn from. So I volunteered. After the first conference on engineering education, I met with the dean and pointed out that the core committee was made up of civil engineering faculty. Well, I'm in chemical engineering so I told the dean, "I will volunteer as the secretary. You can be the chair." The secretary does all the dirty work, and all the terrible work will fall on to the secretary, but I volunteered because that would allow me to suggest who to invite and which speakers to bring in.

So we managed to get Richard Felder and Rebecca Brent to come to our conference. At that time, Malaysia was just changing to outcomes-based education for accreditation, so there were external factors also. When Richard came the first time in 2005 and gave a workshop on outcomes-based education, all the participants were clamoring to get his signature on the paper, and he said, "I felt like a superstar." That was funny.

JW: The other thing that I noticed in the times that I've participated in the conference, you have the ability to draw so many people from so many regions, people from Indonesian, Vietnam, and you're really great at bringing people together from that whole region. How do you do that? How do you get people to travel and connect in this?

KMY: It's mostly regional. To attract faculty from the Philippines, Vietnam, and Indonesia, I will blast emails to them saying who is coming. In the region, when people hear about the keynote speakers that we have, the sessions that we have, they know that the conference will not just be presenting papers. We will also have very impactful workshops that you can really learn from and bring back to your home campus. So every year, I will have some people to be the anchor who will not only come and give keynotes but also give workshops.

JW: So it's important not just to have an international conference but to provide those opportunities for learning that may not be available on their home campus or with experts in the field. I think that is very powerful. So is there a interest in transforming the way engineering education is delivered to students in the region, through outcomes assessment, problem-based learning, which I know that you're very instrumental in promoting?

KMY: Well, outcomes-based education is very big because it's required for accreditation to be a member of the Washington Accord. So in Malaysia, we've done it a long time. We've been a full member since 2009 and a provisional member since 2005. We were provisional members for a while, and back then, only Singapore and Malaysia were members of the Washington Accord. But recently, there's a lot of movement for Indonesia, the Philippines, and India to join, so they are now all jumping on board to get accreditation. So it's not just getting accreditation from another country, but it's their professional body who has to be trained to embrace outcomes-based education.

I'm working in several Indonesian universities now, and they are all changing to outcomes-based education. The outcomes-based education movement is very strong in Indonesia. And because of that, they also have to change the way they teach because it's all about meeting the outcomes and not just a series of topics that you have to teach the student.

JW: When you're working with faculty in Indonesia, you're preparing them for pretty significant changes in how they teach. How do you get them ready? How do you prepare them to be effective change makers?

KMY: When we started a PhD in engineering education program in 2007, that really pushed me to really learn about education research and social science research, which was quite a nightmare. Well, you know how engineers see the world. It's always definite. And when you go into education research, it depends on the paradigm that you take. I say, okay, there's a thing that's called the "research paradigm" in engineering. When I did my PhD, there's no research paradigm. Everything is positive. You get what you see.

I have done studies of faculty members who have gone through the workshops that I conduct, so what actually happened to them? By understanding what happened to them and going through several workshops on change, I'm able to guide those who want to make the change better, to really make sure that they understand that it has to come from the heart. They have to believe in it. If they don't believe in it, it's not going to happen.

JW: Because you know the culture that you're working in, the culture of university life in the region, do you think that makes you just a little more skillful in helping faculty make the move? You talked about using your own research, showing or sharing with them that you've done this yourself. You know the impact that it can have. Do you think there is a cultural requirement to be successful with this particular group or these faculty?

KMY: I think if you want to help people make the change, you have to go through the process yourself, especially things like asking them to change the way they teach. It's easy for administrators and the people in the Ministry of Education to say, "Okay, you need to use this approach. You need to use that approach." But if you have never done it yet, how can you know what is required? So it's not enough to know that something is good. We need to try it first. Then you will know how and why it works. I think making the change is one thing. Sustaining it is another.

JW: Do you want to talk a little bit about that, sustaining the change?

KMY: There are many people who make the change, but then they are not able to sustain their own practice.

JW: And why do you think that is?

KMY: Because it's easy to lose steam, lose the energy, because there will be detractors, people who don't agree with you. And when you do something that is really different like problem-based learning, people tend to confuse this with the normal project-based learning. Problem-based learning, I mean the inductive form like in the medical school application, it's not meant for final year students. Final year students doing capstone projects, that has to be open ended, but problem-based learning is actually very powerful in developing students' self-directed learning skills because you don't teach them first. You give them the problem. You engage them with the problem, and the knowledge is within the context of how it's being applied. So students will see what it's for. Students will see how it's being used. And they don't know how to use it at first, and they need to learn. And as they learn, then they discover it. So you force the students into deep learning, whether they like it or not.

JW: I think that's another challenge that we're talking about in terms of sustaining change. Students, I think, resist at different points in the process. You are proposing a different way to learn. It puts more responsibility on them. They're more actively engaged in their own learning with, like you say, self-directed learning skills. US students are usually not challenged to do something like that, and I think students in other regions are not challenged like that either. So I think that's another contributing factor to sustaining the change.

KMY: Yes, but like I said, you need to believe in it. I have this conviction that for engineers to really be able to contribute, and this will sound like a typical definition, but the purpose of engineering is to make the world a better place. There are a lot of definitions, but that is the one that I hold most because it is true: when you educate engineers well, we have that capability to make the world a better place because engineers are problem solvers, and we are aware of the constraints. We are aware of the consequences of what we do. We're aware of the simplifications that we need to make in order to get things going. But to be able to do that, the world is changing so fast. If you do not have that self-directed learning ability, then how can you keep up with the times with all the changes that are coming?

In problem-based learning, students will also learn how to solve problems together in a team, but the main strength is the self-directed learning part. That is why I go for the inductive vision, rather than the deductive part. So that is what I believe in. If I want to help my students to be good engineers, I need to do this. So because of my own conviction, I am willing to face students' wrath initially, as long as they gain later on.

What I do is I show them the "trauma curve," you know what psychologists term as the trauma curve, consisting of stages of of shock, denial, strong emotion and then finally acceptance, and then they will improve. Once they understand that, then finally they will reach a higher place. So I show my students this trauma curve, and I ask them, "Okay. Where are you?" They think, "This is not happening. How come she's not teaching us? And this is really tough stuff, very difficult engineering stuff. Why is she not teaching us?"

I developed the framework of cooperative problem-based learning because I want to help support the students. This is how we are going to do this class. We go through step by step by step. And I'm here to support you. I'm not throwing you into the deep end and then let you drown because I will be there to make sure that you don't drown. You may swallow some water, but then you'll still be alive. So with the first problem, they will be struggling a little bit. Second problem, still struggling. But by the time they reach the third problem and the fourth problem, I can basically just watch them. I don't have to do anything. I will just be the skeptic and devil's advocate. So what does this actually mean? And if this changed, what will happen. So just those kinds of questions.

JW: Do you see a similar transformation among the faculty as well? This is really changing students. Do you see the same impact on the faculty once they go through this process?

KMY: As a Centre, we run workshops on education research, Scholarship of Teaching and Learning for engineering, for non-education, non-social science faculty members. So we tell them it's not easy. Just because it's education and you've gone through it, it doesn't mean it's going to be easy when you start to have to find the theories. You have read theories and learning to write about what we do, forcing ourselves to reflect and think about how to improve. So we tell them it's not going to be easy, but then if you do this, it will be second nature to you and it'll be easier for you to sustain what you're doing. So I find that those who actually do Scholarship of Teaching and Learning, it's easier to sustain their own practice because the SoTL cycle is like a continuous improvement cycle. They're forced to reflect, they're forced to read what others are doing, and they gain inspiration from others. But those who just write about what they do without really going to the scholarly part, then the likelihood to just abandon their own innovative practice is quite high.

JW: This is a recurring theme in the people I've interviewed for this book. How do we not just make a change, but how do we sustain it? How do we keep moving that forward and supporting faculty as they do that work? So if you were going to give advice to other change makers, faculty who are interested in making change, what are some of the things that you might suggest for them to think about or consider. Is there a helpful hint or something to support them on their way?

KMY: Things will not work your way all the time. Sometimes you seem to be regressing. So this is very much like that trauma cycle. You go the wrong way before you go the right way that you want to go. So you have to understand that human nature will take time. One thing is you need to give it time. You need to be there to support. You need to first be the practitioner and not just expecting people to do things without us doing it ourselves.

We need to lead. We need to lead by example. Have patience. I think you do not have to focus on just one group or one set of places. Sometimes it pays to also go outside of your own university, outside of your own group, because you never know. There's a saying. As a Muslim, we believe that you cannot change other people's hearts. Only God can change other people's hearts. So you do your best, but then God will decide whether it will work or not so it's okay. It is the best. And being scholarly is very important. Whether it works or it doesn't work, write about it so that others can also learn from your experience.

JW: I like your idea that you need to believe in the change. You need to believe in it. It's not merely that I'm just going to put my change idea out there. But having that sense of conviction maybe or belief in what you're doing, I think it's not just great for motivating yourself. Other people see that. They see that you believe in the change that you're trying to make, and you're willing to put yourself forward. You said it's about leading in order to make these changes. I think your example has been very powerful in that regard. You've been leading change in your context and in the region, which I think is really inspiring.

KMY: Well, I'm not perfect. I've made mistakes. Being an engineer, I understand that nobody is going to be perfect, but I think your effort to continuously try is very important.

JW: That's something that we try to instill in our students, the understanding that failure is a part of the engineering process, and you can learn more from your failure than you can from your success. Henry Petroski says success doesn't teach you much of anything. It's the failures that teach you so much and that you can learn from and then apply going forward.

So let me just wrap things up. I wondered if there was some question you were expecting me to ask you that I didn't ask that you'd like to answer.

KMY: I think you've covered everything. Well, okay, I want to offer this. Sometimes people do not accept what we want to do locally. Sometimes even in our own university or even in our own country, it is not something that is normally accepted. Even in Malaysia, although there's movement in engineering education, not many departments will accept that you need somebody in engineering education, unlike I think in the US, where it's more widely accepted.

But then sometimes, you may not see the impact immediately. Sometimes the impact will come later. And sometimes when we do things locally, the impact may be somewhere else. It's good to participate and be part of the community. I think being a part of the community is very important. So not only is it important to make a local community, we should also be part of the international community so that we can see what others are doing. That actually will help us to have the access to a wider worldview and the strength to go on and sustain us in our practice, even at the local level. People always think that the growth should be from the local going to the international, but I found that learning at the international level, I need to bring it to the local level to make the case for acceptance in Malaysia.

JW: Well, it's interesting that you make that point. We met at the conference in Doha, Qatar that included educators from all around the world. It was an amazing experience just because of the breadth of connections that you can start to make. Our connection has dated from that point. But I like that idea that you always expect that you're going to have your biggest impact in your neighborhood, let's say, but you may be connected internationally and have a bigger, more discernable impact away from your own campus.

KMY: If you do not gain that much acceptance locally, don't worry. Just go wherever there is the community.

JW: Right. Find your community. That was one of the reasons I've put this book together, to help people who aren't connected to a community find one in these pages, maybe a community of practice or a community of change makers. Through this book, they'll be connected with other change makers. And I think that can have a profound effect on them and on the changes they want to make.

Index

Page numbers in *italic* indicate figures.

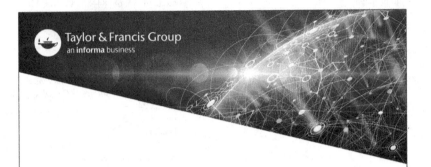

Printed in the United States
by Baker & Taylor Publisher Services